Good Crop / Bad Crop

Good Crop / Bad Crop

Seed Politics and the
Future of Food in Canada

Devlin Kuyek

Between the Lines
Toronto

Good Crop / Bad Crop

First published in Canada in 2007 by
Between the Lines
720 Bathurst Street, Suite #404
Toronto, Ontario
Canada M5S 2R4

1-800-718-7201

www.btlbooks.com

Kuyek, Devlin, 1973-
        Good crop/bad crop : seed politics and the future of food in Canada / by Devlin Kuyek.
Includes bibliographical references and index.
ISBN 978–1-897071–21–2

1. Seed industry and trade – Economic aspects – Canada. 2. Seed industry and trade – Government policy – Canada. 3. Seed industry and trade – Canada – History. 4. Plant-breeding – Canada. 5. Plant varieties – Protection.
I. Title.
HD9019.S432C3 2007   338.1'70971   C2007–906277–6

Cover design by Jennifer Tiberio
Text design and page preparation by Steve Izma
Printed in Canada

Between the Lines gratefully acknowledges assistance for its publishing activities from the Canada Council for the Arts, the Ontario Arts Council, the Government of Ontario through the Ontario Book Publishers Tax Credit program and through the Ontario Book Initiative, and the Government of Canada through the Book Publishing Industry Development Program.

Canada Council   Conseil des Arts
for the Arts   du Canada

Canadä

ONTARIO ARTS COUNCIL
CONSEIL DES ARTS DE L'ONTARIO

For Pete's sake.

Contents

Acknowledgements

THIS BOOK IS MY ATTEMPT to share what I have learned from the wise and inspiring farmers, researchers, and activists I have met as the writing process unfolded.

There is a long list of people who have helped directly with this book and its various versions. Among others, I wish to thank Brewster Kneen, Louise Vandelac, Harriet Friedmann, Birgit Müller, Terry Boehm, Andrew Skinner, Dominique Caouette, and my colleagues at GRAIN.

I am deeply grateful to Paul Eprile and Robert Clarke of Between the Lines for their fine editing work, and to everyone at Between the Lines for helping to bring this project to fulfilment.

Thank you always to Alisha, Una, Fanon, and the rest of my extended family.

Part I

Roots

Chapter I

Transformation

A FARMER SOWS SEEDS. Seeds grow into plants. The plants are harvested and some of the seeds are returned to the earth to produce another crop. This cycle is the foundation of agriculture. While it may appear simple, there is tremendous complexity within. The new seeds are always slightly different from the old, just as a farmer's field is always different from season to season: the climate changes, diseases and pests come and go, and rainfall varies. Through the variation of their seeds the plants enable future generations to adapt to their surroundings, ensuring the survival of the species.

Agricultural plants do not carry out this evolutionary process alone. People, most often farmers, can and have always encouraged and shaped it by selecting and replanting seeds from those plants that fare best in their fields or satisfy certain cultural interests. They have also intervened more directly by deliberately crossing varieties to breed plants for the attributes they desire. The world of plant breeding consists mainly of these processes of selection and crossbreeding, and today's wealth of agricultural biodiversity is a result of generations of plant breeding efforts on the part of farmers and, more recently, scientists.

The seeds we plant are profoundly social: they reflect and reproduce the cultural values and social interests of those who developed them.

When they were the exclusive domain of farmers, seed systems were characterized by diversity – a kaleidoscope reflecting the many hands nurturing the seeds, the unique territories where they were planted, and the tastes of the many mouths that enjoyed the results. But during the twentieth century, much of the work with seeds shifted into centralized plant breeding programs run by scientists. These programs, with the backing of governments, focused on the development of a few varieties that could be used over large areas. Crop diversity narrowed immediately, while the potential for a single plant variety to transform farming and food systems on a large scale increased exponentially. Seeds became vehicles that could be used for deliberate social and political transformation.

The Green Revolution of the 1960s and 1970s is probably the most well known and dramatic example of how seeds can be used to such ends. Under a U.S. Cold War program to increase food production and thereby stem the spread of communism, scientists developed a few high-yielding varieties of major cereal crops and deployed them throughout Asia and Latin America. These Green Revolution varieties did increase yields, and proponents pointed to the rapid shift from grain shortages to surpluses in certain countries where the Green Revolution model was put into practice. But the yield increases were only possible under specific conditions – and those conditions required irrigation and the intensive use of chemical fertilizers and pesticides. These varieties created new markets for agribusiness, both for inputs (fertilizers and pesticides) and for the grain trade, because of the resulting surplus production of a few cereal crops.

The benefits for farmers weren't so clear-cut. The new varieties brought both environmental and health calamities, such as soil erosion, nitration, and pesticide poisonings, and socio-economic problems. Many poor farmers, either unable to afford the inputs required or farming on lands not suited to the plant varieties, could not compete, and were forced to leave their lands and look for work in the rapidly swelling cities.[1]

Back in 2000, I had my first of several face-to-face experiences with the Green Revolution. I was travelling with a peasant organization in the Philippines. We were visiting farmers in Isabela province towards the north of the Philippines archipelago, in a corn-farming area. The local corn varieties were abandoned in the 1970s to make way for "high-yielding" hybrid corn seeds that the government promoted as part of a credit package. Unlike the local varieties, the seeds saved from the hybrid plant do not grow properly, so the farmers were forced to purchase new seed for every new season. Almost all of this corn seed is produced by two U.S. pesticide companies, Monsanto and DuPont, and sold by a few local merchants who monopolize the grain trade in the area and sell the seeds as part of a package along with pesticides and chemical fertilizers.

The local farmers told us that they pay the merchants an incredible 30 to 40 percent interest per cropping season (60 to 80 percent per year) and are forced to sell their harvests back to them for whatever price the merchants offer – which is invariably low, fluctuating between U.S. 8–16¢ per kilogram. The peasants, as a result, live and work within a vicious cycle of debt and dependency. They grow hybrid corn because that is the only crop they can sell, and they need the money to pay off their debts. But in the process they generate further debt. They also lose their capacity to grow food for themselves. The hybrid corn is good only for animal feed, so farmers have to buy what food and household goods they can afford. Many of their children suffer from malnourishment. One older peasant woman, sitting in the midst of corn grains she was drying, turned to me and said flat out, "I want to die."

A few years later, in 2004, I had a completely different experience. I was in a small village in Bangladesh, close to the city of Tangail, with some colleagues. We were meeting with women from a grassroots movement of farmers, known as *Nayakrishi Andolon*. Out of their disenchantment with the Green Revolution programs pushed in the country they had decided to abandon the high-yielding varieties and chemical inputs to return to their traditional varieties and practices. The movement had by then grown to include around a hundred thousand farmers.

The farmers took us to see their local Community Seed Wealth Centre. Here the seeds of hundreds of different varieties of dozens of different crops were stored in a bewildering number of clay pots and glass bottles. A farmer explained that this collection was only a small part of a larger seed system linking hundreds of communities throughout the country in a sophisticated exchange and monitoring network that ensured that at any point in time thousands of different seed varieties were being grown and kept alive, somewhere. Each variety had a name, and the farmers described their characteristics – some known for their taste, others for their yields, others for their resistance to pests and diseases. The conversation eventually turned to the idea of food sovereignty – a slogan that has emerged from peasant and indigenous peoples' movements around the world, as a way to describe the multi-dimensional importance of the agriculture they practice, and to unify their struggles against the common threat of industrial agriculture. When someone asked the women farmers what food sovereignty meant to them, one pointed to the seed centre behind her, smiled, and simply said, "This."[2]

The Nayakrishi farmers told us how the loss of seed from a household means the loss of women's power in that household. A dependence on the outside market for seeds would make the women redundant, displacing them from the control of what lies at the heart of their food systems. They also understood that seeds from outside can carry their own agendas, bringing in conditions that are alien to local farming practices and culture. Having lived through the Green Revolution, they knew all too well that industrial agriculture comes in by way of the seeds and that today, especially with the advent of genetically engineered crops, seeds can be used to impose a model of agriculture where peasant farmers have no place and where a small number of global corporations control the entire food chain. For them, food sovereignty started with keeping control of their seeds in their own hands.

North America's Green Revolution

Across the ocean, in North America, seeds have also played a major role in transforming farming and food systems. The high-tech farm machinery and chemical fertilizers and pesticides of today's industrial agriculture could only be adopted on North American farms after the development and widespread distribution of suitable, standardized varieties. And it was through this large-scale farming of a few varieties that the big, grain-trading companies emerged and soon came to dominate the North American and, later, the international food trade.

Although these transformations of agriculture in North America laid the foundation for agribusiness, they emerged out of public projects. It was public plant breeding programs that by and large developed and supplied the mass distribution of seeds. The initial public programs relied on open systems of plant breeding based on collective research and the free flow of seeds. Seeds were viewed more as a public good than as a commodity, and the breeding objectives, defined according to perceived national interests, were widely shared by scientists, governments, agribusiness, and farmers. Today the public seed systems of North America, like those in other parts of the world, are in steep decline. A radical transition is afoot through which a few transnational corporations, with the active participation of governments, are taking full control over seeds.

The Rise of Transnational Seed Corporations

There is money to be made from seeds. To be specific, the world's seed markets are worth around U.S. $22 billion, and rising. While this is not an insignificant sum in its own right, seeds are actually worth much more than their sticker price to corporations. Unlike other agricultural inputs, seeds, thanks in large part to the Green Revolution, contain the extraordinary potential to allow corporations to determine farming practices, and to enhance corporate power within the entire food chain.

Monsanto, a U.S.-based pesticide and pharmaceuticals corporation, has used this capacity to great effect. In the mid-1990s, while it was in the midst of an unprecedented wave of takeovers of seed companies, it commercialized genetically modified (GM) soybean, cotton, canola, and corn seeds that were designed to tolerate the application of its blockbuster herbicide, glyphosate (tradename Roundup). By 2005 these GM seeds were sown on over 180 million acres worldwide, driving sales of glyphosate through the roof, and making Monsanto the world's largest seed and fifth largest pesticide company.[3] Other pesticide companies soon followed Monsanto's lead. Between 1997 and 1999 alone, transactions by pesticide companies in the seed industry topped U.S. $18 billion – not far off the value of the entire global, commercial seed market.[4]

Pesticide corporations had a clear agenda for investing in seeds. By the 1980s the pace of discovery of new chemical formulations had slowed, and it was becoming increasingly difficult and expensive to identify and develop new pesticides. At the same time several of the blockbuster pesticides, such as glyphosate, were set to come off patent, and the larger corporations feared that competition from generic producers would lower prices and reduce their market share. Off-patent pesticides were expected to account for 69 per cent of the entire pesticide market within a couple of decades.[5] Plant breeding, and in particular the newly developed technique of genetically engineering plants, offered a solution to these looming problems. They addressed the difficulties of finding new pesticides by opening up biology as a whole new area of science in which the industry could identify and patent new pesticide technologies, this time based on the DNA of organisms. Some 26 percent of the GM crops planted in the world in 2003 were engineered with genes from *bacillus thuringiensis* (Bt), a soil microbe that is toxic to certain crop pests. The technology is most often used as a substitute for certain insecticides, and is generally integrated within a regime of pesticide applications, without any significant changes to agricultural practice.[6] Pesticide companies could also use biotechnology to counter generic competition by genetically engineering plants for dependence on their brand-name pesticides, as Monsanto did with glyphosate. GM crops were expected to involve lower regulatory costs too; a new pesticide cost a company between U.S. $40 million and U.S. $100 million to bring through the regulatory process, whereas it cost less than U.S. $1 million to bring a new conventional plant variety to market.[7]

Table 1
Global Pesticide and Seed Sales and
Rank of Top Companies – 2005

Company	Rank: Pesticides	Sales: Pesticides (U.S. $millions)	Rank: Seeds	Sales: Seeds (U.S. $millions)
Bayer	1	$6,120	8	$387
Syngenta	2	$6,030	3	$1,239
BASF	3	$4,141	–	–
Dow	4	$3,368	–	–
Monsanto	5	$3,180	1	$3,118
DuPont	6	$2,211	2	$2,600

Source: ETC Group

Corporate interest in seeds has by now, however, gone far beyond a simple concern with selling seeds and pesticides. With the advent of genetic engineering and the expanding scope of patents and other monopoly rights, seeds have become one of the key points of proprietary control in the global food order; those who control the seeds have an increasing amount of leverage over other sectors, be they farmers, food processors, or retailers. Control over seeds has even become critical to the latest corporate scramble for the global agrofuels market, with companies like British Petroleum now investing heavily in seed genetics. The inevitable result is more vertical integration, with a tendency towards fully integrated supply chains from seeds to the supermarket shelf (or to the gas station), in which the "market" is only present at the end of the food chain. Robert Fraley, executive vice-president of Monsanto, puts it this way: "What you're seeing is not just a consolidation of seed companies, it's really a consolidation of the entire food chain. Companies like ours, who want to continue to be in the food and feed production business, are all trying to secure our spot along that chain."[8]

Corporate investment and consolidation in the seed industry is only going to intensify, both in terms of buyouts and mergers within the industry and in terms of vertical integration with downstream agribusiness and agrofuel corporations. Already by 2002 ten seed companies controlled upwards of a third of the global commercial seed market, valued at U.S. $23 billion, with four companies controlling 86 percent of commercial corn seed and 49 percent of commercial soybean seed. In 2006 and 2007, there was another flurry of mergers and buy-outs. Monsanto bought up the world's largest vegetable seed company, Seminis, and the world's largest cotton seed company, Delta & Pine Land, and announced a joint venture on biotechnology research with fellow pesticide and seed corporation BASF. Its major competitors, DuPont and Syngenta, also announced a joint venture for their seed operations. Going into 2007 the top ten companies had increased their control over the global commercial seed market to approximately 55 percent.[9]

These are alarming developments. In Canada, we have already seen how pesticide companies have used their control over seeds to recklessly contaminate our food supply with transgenic material that has not been proven to be safe and that offers no benefits to the average Canadian. Now these same companies are capitalizing on the fears they have instilled in consumers and farmers, by designing systems to "manage" the risks that they created. These new systems will give them greater control and crush anything outside of the agribusiness box, even, or especially, those traditional seed practices that people have used safely for generations and generations. The coming introduction of seeds genetically engineered to produce pharmaceuticals will take this process to new heights. For the first time, crops that cannot enter our food supply will be grown on a large scale. Contamination is inevitably bound to happen, and when it does, seeds will forever more become hazards, risks that must be sealed off in fully integrated, sanitized, corporate chains of production to guarantee "safe" food – for those who can afford it. The rest of us will have to live with the contaminated, bulk product churned out by industrial agriculture. Any effort to produce food outside of the system will be condemned as a risk and heavily persecuted. A grim scenario, but we are already quite far down this path: the infrastructure is there, just waiting for the next crisis to bring it into full operation.

Complicated Commodities

Thankfully, seeds have a certain inbuilt resistance to corporate control. To the great frustration of seed companies, seeds reproduce themselves. Out of one seed, a farmer can produce ten more, and eventually keep his or her fields full without ever needing to return to an outside supply. So, historically, seeds have not been sold. Farmers have saved them from their harvests or exchanged them with neighbours. Even today, after all the corporate investment in the seed industry, farm-saved seed still supplies the bulk of seed planted around the world, particularly in developing countries where farmers directly supply about 70 percent of their own seed needs. And the situation is not so different in many industrialized countries: 95 percent of the cereal seed sown in Australia is farm-saved, while in Canada farm-saved seed accounts for 76 percent of the wheat seed supply, 66 percent of the barley seed supply, and 74 percent of the seed supply for peas.[10]

The other problem for seed corporations is that seeds are collective, in that each seed is a product of generations of selection or plant breeding by farmers and scientists. Any new variety is just a slight modification of previous types, making it difficult for any one company to claim ownership of a plant variety, or to stop others from using a given variety for multiplication or further breeding work. Not such a good situation for seed corporations, but an ideal one for public-style breeding programs with their open culture of co-operation among scientists.

Seeds, therefore, have never provided fertile territory for corporate profits, and when corporate interest in seeds surged in the 1980s and 1990s there were no ready-made markets for them to step into. The markets had to be built; a new architecture of laws and regulations had to be erected to make it possible for corporations to manage these two long-standing constraints to their profits. Against all historical precedents, seeds had to be turned into commodities.

This book is about how this process has unfolded and is still unfolding in Canada. In this country, seed corporations, with the support of governments, have moved aggressively to stop seed-saving and to privatize our seed system by way of a steady stream of laws, regulations, and technologies. Through it all, farmers, gardeners, public plant breeders, and citizens in general have been shut out of any meaningful participation. The age-old act of saving and planting seed – the very basis of human civilization, is now wrapped up in a complexity of patent rights, contracts, and deep risks to the environment and human health. These obstacles make it increasingly difficult for anyone without a good lawyer to engage in the act. It has gone so far that we are now at the point where seed-saving is actually treated as a crime, farmers can go to jail for it, and what is worse, Canadians are starting to see this as normal.

My simple objective is to cast some light on this process, to unpack what has happened, and to put it into a historical context that provides some perspective on the enormous transition that has taken place over the past hundred years. The book focuses on Canada, where corporate control over seeds is highly advanced, but what is happening here sets an ominous example for what is to come in the rest of the world. So this book should be of use to people outside Canada as well.

I also hope that readers will come away with a better understanding of how the construction of a corporate seed system is destructive of other types of systems, be they public, such as that which defined Canadian agriculture in the twentieth century, or the farmer systems that continue to serve the food needs of much of the world. These remain the foundations for any challenge to corporate control. Control has passed from farmers, to the state, and now to corporations. Many Canadians are uncomfortable with this latest move, and some are actively resisting it. But unknown to most people is the capacity that corporations acquire, through control over seeds, to impose a single, integrated food system upon us, a system essentially designed to meet the needs of a small number of transnational agribusiness and food corporations. If we want something different, and there is every reason to want it, then we are going to have to find a way out. My hope is that this book will provide a little light and stimulus for the journey.

Chapter 2

Industrialization

IN MARCH 2007 I was at an agriculture exhibi-
tion outside of Buenos Aires, Argentina, where Latin America's
new model of agribusiness was on display. Spread out over a few
hundred acres were rows of multimillion dollar tractors and pesticide sprayers, sample trials of new genetically modified plants, the
latest 4x4 trucks, and, of course, banks offering special incentives
to those tempted by the glitzy new equipment. It was a big-farm
shopping mall in the middle of the Third World, selling the same
illusions of modernity and wealth as any big box store in a North
American suburb.

Oddly absent from the "agroexpo" were farmers. I met a few
here and there, but most of those scoping out the high-tech
equipment were businessmen from the city. And, as I thought
about it, I didn't remember having seen any farmers as we drove
past kilometre after kilometre of genetically engineered soybean
fields on the way to the agroexpo. Farmers, at least as I've always
understood the word, didn't appear to be part of Argentina's new
agribusiness fiesta.

"Argentina's agriculture is an agriculture without lands, without
labour, without capital, and only with knowledge." So said Gustavo
Grobocopatel, head of Los Grobo, Argentina's largest soya producer, to a reporter with Agence France Presse, in June 2007. He
proudly explained that around 70 percent of the lands used for soya
production in the country is rented and 80 percent of the farm
work, such as seeding and harvesting, is subcontracted. Everything

is simplified and mechanized to the extreme, thanks in part to the widespread use of genetically engineered soybeans. These make weed control an easy matter of spraying herbicides at planned-out intervals. "Our company used to have a hundred people in the fields and ten in the offices; today we are ten in the fields and a hundred in the offices," he added.[1]

Back in Canada, agribusiness executives rarely talk so boldly of a future without farmers. Indeed, their agribusiness-speak almost always centres around farmers and the somewhat mythical "farm family." Although the average farm is nothing like it was a hundred years ago, it is true that most Canadian farms are still run by farmers who live on the land.

Why should agribusiness companies care about taking over Canadian farms anyways? It is not as if this presence of farmers has been much of an impediment to corporate profits. Consider the following: While total gross revenues for Canadian farmers doubled between the late 1940s and the end of the twentieth century from $17 billion to $35 billion, the total net farm incomes fell over the same period from $11 billion to under $5 billion. Meaning that, as Canada's National Farmers' Union likes to point out, while Canadian farmers used new inputs and other technologies to increase their production by about $18 billion, more than doubling it in those sixty years, the corporations that sold them those technologies not only swallowed up the entire $18 billion in increased production revenue, but took an additional $8 billion from farmers' pockets as well![2] My retort to leaders of farm organizations when they roll out such figures is always, "So how have farmers let agribusiness get away with such obscene profit-taking, especially when farmers have to take on all the risks of running the farm, the unpredictable weather, the volatile markets?"

I have no illusion that there is an easy answer to this question. It runs far deeper than farmers simply being duped. Richard Levins and Richard Lewontin, professors at Harvard University, seem to think so too. They suggest that we can best understand how this has come to be by first drawing a distinction between agriculture and farming. As they put it,

> The basic problem in analysing capitalist development in agriculture is the confusion between farming and agriculture. Farming is the process of turning seed, fertilizer, pesticides and water into cattle, potatoes, corn and cotton by using land, machinery and human labor on the farm. Agriculture includes farming, but it also includes all of the processes that go into making, transporting and selling the seed, machinery and chemicals used by the farmer and all of the transportation, food processing and selling that go on from the moment a potato leaves the farm until the moment it enters the consumer's mouth as a potato chip.[3]

Put more simply, farming is what is carried out on the farm; agriculture includes farming and all the activities that both occur upstream from the farm (the production of what is used on the farm – tractors, fertilizers, pesticides) and downstream from the farm (the use of what is produced on the farm – trade, processing, retail sales).

In Canada, as in other parts of the world, instead of displacing farmers and taking over farms directly, companies have maximized their profits and taken control over agriculture by continuously displacing agriculture from the farm into non-farm, industrial production processes. Tractors built on assembly lines replace human and animal traction; synthetic fertilizers produced in factories replace organic fertilizers and nutrient cycling practices; chemical pesticides replace traditional pest management practices. Farmers, in most cases, still own and operate the farms and do the farming, but they are gradually separated from the means of production that they now purchase as commodities (seeds, fertilizers, fuel, tractors).[4] As the farm becomes more dependent on these outside inputs the dynamic interaction between farmers and their lands, soil, livestock, and environments breaks down and farms come to resemble the factories and structures that supply them, while the farmer comes to resemble an industrial labourer or a low-ranking corporate manager.

Some people think that farmers have little choice but to go along. They argue that when a new technology that increases productivity comes around, there is always an initial set of farmers, the so-called "early-bird" or "progressive" farmers, that adopts it in the hopes of increasing profits. When more farmers adopt the same technology and increase their productivity, supplies of their crops go up and drive down the market price. The drop in prices puts pressure on other farmers to increase their productivity, leading more and more farmers to adopt the technology, and crop prices fall further and further. U.S. agronomist Willard Cochrane calls it the treadmill effect:

> The aggressive, innovative farmer is on a treadmill with regard to the adoption of new and improved technologies on his farm. As he rushes to adopt a new and improved technology when it first becomes available, he at first reaps a gain. But, as others after him run to adopt the technology, the treadmill speeds up and grinds out an increased supply of the product. The increased supply of the product drives the price of the product down to where the early adopter and all his fellow adopters are back in a no-profit situation. Farm technological advance in a free market situation forces the participants to run on a treadmill.[5]

In Cochrane's scenario the big losers are the "laggard farmers" who do not adopt new technologies and who must eventually sell their farms or go bankrupt. "Progress" or "rationalization" in agriculture, he argues, may be "cannibalistic," but it is unavoidable.

One problem with the treadmill idea is that it assumes a free-market situation that has rarely, if ever, existed in North America. Local markets for farm produce have generally been dominated by a few "middlemen" who sell inputs and purchase outputs.

Through their near monopolistic positions, these companies exert significant control over not only crop prices, but also farming practices. Governments too have actively shaped farmers' choices. In the post-World War II period, governments in North America intervened directly into agriculture with policies and regulations that prioritized the development of large-scale, globally competitive agribusiness and discouraged small-scale farming. Agricultural modernization has accordingly been directed or pushed down the path of "rationalization" – specialization, farm consolidation, crop uniformity, and other techniques of "efficiency" that encourage the adoption of new, industrial technologies.[6]

The treadmill idea also makes it easy to forget that technologies emerge within a historical context that frames farmers' choices. There is no reason to assume that increased productivity is inherently tied to industrial inputs. Many existing alternative techniques and technologies could be pursued to increase productivity in non-industrial ways.[7] Yet agricultural research in North America has focused almost exclusively on the development of technologies and practices that reinforce industrial agriculture. After World War II, when the treadmill really took off, decisions over agriculture research policy were in the hands of small, homogeneous circles of male scientists, company representatives, and some farmers, united by a desire to use modern science to increase agricultural production of major commodity crops for the good of the nation. Any associated social and environmental costs were by and large ignored, as were the possible contradictions among the different actors, such as those between farmers and agribusiness.[8]

Few farmers were actually involved in the decision-making processes for agricultural research, and those few shared the prevailing assumptions about agricultural progress. Agricultural research was mostly a one-way process, with knowledge and technologies flowing from the labs to the farms. Often the seeds were the messengers. Productionist agriculture was embedded in the seeds developed and released by the centralized plant breeding programs of the time. It was largely through seeds, then, that agriculture was displaced from the farm and integrated into the machinery of agribusiness. Industrial agriculture had emerged.

Over time, however, the nation-building vision that held together the post-war agriculture coalition faded away. The infrastructure of the public seed system was dismantled. Today, seed politics are at the centre of another transformation, away from the postwar national agriculture research systems, to transnational structures directly managed by agribusiness corporations. Corporations are taking industrial agriculture to new extremes, and, in the process, robbing the growing backlash against industrial agriculture of its basic resource – the seeds.

Efforts to build alternatives to industrial agriculture must now diversify the seed system, putting seeds once again in the hands of farmers and democratizing the space for seed politics. There is, at root, a basic contradiction between the two models: with industrial agriculture the farm adapts to the seeds; with its alternatives, such as small-scale organic agriculture, seeds are adapted to the farm. The problem for the latter, however, is that today the very mechanisms that corporations use to solidify their control over seeds and food systems preclude such adaptation. You cannot adapt plants to your farm if you cannot select and save seeds from those plants for next year's planting. Likewise, you cannot build alternatives to industrial agriculture when your seed supply is entirely dependent on transnational corporations whose very business is predicated on the growth of industrial agriculture. It is a simple problem with no simple solution.

In Canada seeds were taken out of farmers' hands a couple of generations ago and today it is hard to imagine them going back. It is even harder to imagine them residing in the same number of hands, given the decline in the farm population. But it is harder still to imagine leaving seeds in the hands of agribusiness. Fortunately, corporate control over seeds is not complete and there are many ways left for us to resist it. First we will need to understand the processes through which industrial agriculture turns seeds into commodities. And we will have to identify and support the many spaces where alternative seed systems can be explored. Collectively we can begin to build anew. We can take the best from the rich history we have in this country of public plant breeding, farmer seed systems, and indigenous knowledge. The future may not be so bleak after all.

Part II

A History of Seed
Politics in Canada

Chapter 3

Germination

Before the arrival of Europeans, First Nations peoples in the lands now known as Canada managed highly developed systems of agriculture and permaculture, using seeds from a variety of crops – squash, maize, sunflower, and beans, to name a few – that they had carefully selected for and nurtured over generations. European colonization destroyed much of this agricultural diversity – the plants, the animals, the knowledge – but a significant amount of integration took place with settler agriculture. The seeds that the initial waves of immigrants brought were not suited to the new climate, short seasons, unfamiliar soil types, and parasites. They rarely led to decent crops. The settlements were able to survive only by adopting some of the First Nations peoples' crops, particularly corn.[1]

Unlike the situation in the U.S., where early colonial agricultural policy sought to develop export plantations, in Canada the mercantile system was built on fur and fish exports, with little in the way of agricultural exports. Agriculture served the food requirements of the people living in the colonies, and the seed systems that developed during this period focused on the creation of an agricultural diversity adapted to both Canadian conditions and the cultural demands of the settlers. Seeds were in the hands of the European farmers and gardeners, and over time, through their networks of farmer-to-farmer exchange of agricultural knowledge and seeds, the new Canadians built a solid foundation of agricultural diversity.[2]

It was only in the latter part of the nineteenth century that governments started to play a more active role. In 1885 Professor William Saunders was commissioned by the Canadian Parliament to look into the experimental farms that the United States government had set up to boost agricultural research. Saunders was pleased with what he saw in the U.S. and in his report to Parliament he called for the establishment of similar farms in Canada. In the following year, five federal experimental farms were established, with another twenty opening up between 1905 and 1916. In 1888 Saunders was named the first Director of the Dominion Experimental Farm in Ottawa.[3]

One of the priorities for these farms was to support the improvement of seeds. But at this point, "improvement" was not confined to the experimental stations. They were only a part of a larger process of innovation still carried out primarily by farmers. Saunders and his assistants focused on extensive collection missions to gather seeds from farmers' fields around the world. These were then multiplied on the experimental farms and sent out to farmers for further experimentation and selection. There was no shortage of interest among Canadian farmers. In 1895, the year Saunders began sending out packets of free seed varieties to farmers, he was swamped with 31,000 requests for seeds and was only able to send out 26,000. The next year, over 35,000 packets were sent out and the program continued at this rate until the end of the nineteenth century.[4]

This initial, public seed program reflected the government's awareness of the willingness and capacity of farmers to experiment with, and improve the performance of, different varieties of seeds. There was little reason for the state to think otherwise. Farmers led plant breeding and agricultural research, in general, and by bringing varieties from their mother countries, exchanging seeds widely among themselves, and continuously selecting from within their crops, they produced significant results in a relatively short time.

The development of Red Fife wheat is probably the most widely known example of the strength of this farmer-to-farmer system. Wheat, the principal crop of the European diet, had proved particularly difficult to adapt to Canadian conditions because of the short growing season and the European varieties' susceptibility to rust. At the time of Confederation, farmers were still having difficulty producing enough wheat to supply local markets, especially in the prairie settlements to the west where all efforts to grow wheat had so far failed. The situation changed dramatically in mid-century after David Fife, a Scottish farmer in what was then Upper Canada, planted seeds of a variety that would become known as Red Fife. Fife received the wheat seeds from a friend in Glasgow who had collected them from a ship sailing from Poland carrying wheat from the Ukraine. Red Fife had good resistance to rust, and, most importantly, it matured early enough to avoid the frost. It was also ideal for bread-making. From David Fife's farm near Peterborough, seeds of Red Fife spread rapidly from farmer to farmer across North America.[5]

Chapter 4

Nation

RED FIFE WAS A TESTAMENT to the strength
of Canada's farmer-centred seed system, but, ironically, also a
harbinger of its demise. The variety made wheat production on
the Prairies a possibility, and, as a result, it immediately caught
the interest of industry and the Canadian government. The variety
had the potential to help the federal government settle the West,
and its excellent milling qualities were quickly recognized by
those wanting to turn the Prairies into the "Grain Elevator of the
British Empire." By the end of the nineteenth century the Canadian Pacific Railway and the Hudson's Bay Company were already
holding fairs in the Western settlements and establishing their
own experimental farms to promote Red Fife. The federal government introduced a series of measures to support its adoption.[1]

Still, William Saunders, the director of Canada's experimental
farm system, was not entirely satisfied with the performance of
Red Fife, and discussions with farmers and industry people convinced him that something better needed to be found, particularly
when it came to maturity time. He had other varieties in his collection that matured earlier, but none of them had the superior
milling quality of Red Fife. So, instead of searching the world for
new varieties as he had previously done, Saunders and his team
embarked on a seminal change of course: they began to try and
"improve" Red Fife itself.

Saunders was taking his cues from a development sweeping
across the continent's agricultural research institutions. In his

early days, Saunders and the other "gentleman" plant breeders of his era followed a Darwinian approach, sometimes called gradualism. They selected plants from a large population, randomly cross-pollinated them, and then repeated the process with the progeny. Each generation would improve slightly on the former.[2] Their methods, despite their efforts to cloak their work in professionalism, were in reality little more than a dressed-up version of the traditional selection practices of farmers. They could be refined and taken up by scientists, but they could also be carried out by, or in collaboration with, farmers on farms. Indeed, such methods were most effective when undertaken by many people working in many locations. This was the case with the federal seed mail-out program.

In 1900, however, three European scientists working separately (Hugo de Vries of the Netherlands, Eric von Tshermak of Austria, and Carl Correns of Germany) simultaneously rediscovered Mendel's theories on hard inheritance. Six years later the British scientist R.H. Biffen applied the theory to plant breeding, demonstrating that disease resistance was inherited by "factors of inheritance" – later known as "genes." And, shortly after, the Danish scientist W.L. Johanssen developed the "pure line" method to "fix" desired genes into cultivars. Here, finally, was a means of distinguishing plant breeding from farming, and of conferring upon it the status and legitimacy of a Science (with a capital "S").[3]

William Bateson, the founder of the term "genetics," epitomized the new triumphal spirit in a speech to a conference of plant breeders in New York in 1902:

> [The plant breeder] will be able to do what he wants to do instead of merely what happens to turn up. . . . The period of confusion is passing away, and we have at length a basis from which to attack that mystery such as we could scarcely have hoped two years ago would be discovered in our time. . . . The plant breeder's new conception of varieties as plastic groups must replace the old idea of

fixed forms of chance origin which has long been a bar to progress.[4]

Plant breeding was quickly converted into a science of precise laws, methods, and techniques, such as pedigree breeding and gene transfer. The methods were beyond the capacity of most farmers, so they were shut out of the plant breeding process. So too were their farms. While gradualism focused on quantitative traits, like yield, that plants possess in various degrees and that are highly influenced by external factors, Mendelian breeding was based on qualitative traits controlled by single genes that are either present or absent in plants. Breeders could therefore focus on "fixing" plants with certain genes, such as the one for disease resistance, without having to worry about adapting them to the specific, local conditions of farms. There was every reason, then, to take plant breeding off the farm and into the controlled environment of scientific research stations, under the central command of scientists and technocrats.[5]

Saunders wasted no time in applying the new Mendelian science to Red Fife. He began crossbreeding it with varieties in his collection that matured earlier, and soon his efforts paid off with the development of Markham wheat, a cross between Red Fife and an Indian variety called Hard Red Calcutta. His son Charles, a devotee of the Mendelian school, who in 1903 became the head grains researcher at the Ottawa experimental farm, later selected a cultivar from within a population of Markham that performed particularly well. Samples of this variety, called Marquis, were sent out to Prairie farmers in 1909. By 1920, the Marquis strain accounted for 90 percent of the hard, red, spring wheat on the Canadian Prairies. It was sown on more than 20 million acres in North America, from southern Nebraska to northern Saskatchewan.[6]

The overwhelming popularity of Marquis was not simply a matter of its performance in the field. The variety was introduced in the context of a federal program to rapidly settle the

Prairies and increase wheat exports. Many of those who began farming the Prairies at the time and who took up Marquis wheat were either recent immigrants, without access to adapted seeds, or speculators interested in making a rapid return on investment. Moreover, state policies and high prices for wheat, particularly during World War I, when Canada was called upon to meet the flour supply needs of its allies, encouraged farmers to abandon cattle, hogs, poultry, and gardens to focus on growing wheat.[7]

The expansion in wheat acreage required an expansion of the seed supply, and the gap was filled, not with a massive distribution of seed packets containing different varieties, but through the multiplication and distribution of Marquis wheat. With the passage of the *Canadian Grains Act* in 1912, Marquis became the standard against which all other wheats were measured.[8]

In the archives of the Agriculture and Agri-food Canada (AAFC) library in Ottawa there is a meticulous study of Marquis wheat that Stephan Symko, a Ukrainian-Canadian plant breeder with AAFC, researched and wrote in 1999, just before he died. He points out that Marquis was not the only option available to farmers at the time. Seager Wheeler, a farmer from Rosthern, Saskatchewan, well known for his plant breeding work, had selected a red wheat variety from his fields that appears to have been a natural hybrid of the Red Fife, Preston, and White Bobs varieties he was growing. This variety, known as Red Bobs, had a better milling quality than Marquis and became one of the principal varieties grown in southern Alberta, where it thrived. Another strain was the Kubanka variety, most likely brought over by Ukrainian immigrant farmers. It not only produced an excellent flour quality but, with its resistance to rust and its very strong awns (protective bristles) and root systems, was well adapted to prairie conditions. Unfortunately, Symko never tells us why these other varieties were cast aside and forgotten in the haste to make Marquis the dominant crop. Greater diversity could have prevented a lot of hardship.[9]

The immediate results of the narrow focus on Marquis were catastrophic. In 1916 Marquis's resistance to wheat stem rust broke down and nearly a third of the total harvest was lost. Charles Saunders tried in vain to identify and incorporate new resistance genes into Marquis without reducing its milling quality. Frustrated, he began to wonder "whether the discovery of Mendelian unit characters is sometimes due to the unhappy combination of a great deal of enthusiasm with very few facts."[10] Furthermore, by 1920 wheat monoculture was already producing widespread soil erosion. The problem worsened during the 1920s, leading eventually to massive crop failures. The crisis was compounded by a dramatic plunge in global wheat prices.[11]

Even so, the hard experiences with wheat in these early years did not challenge the dominant model of agricultural development pursued by the Canadian government. During the 1920s and 1930s, largely in response to the demands of an increasingly organized and vocal farm population, Ottawa established marketing boards, pools, and price support mechanisms to support the production of wheat and other commodities. With the exigencies of World War II, the government made further interventions, building up an extensive machinery to control production, prices, and exports of basic commodities, including foodstuffs. Those agricultural subsidy, support, and stabilization programs that were implemented as part of the war effort endured long after, due to the political backing from both farmers and industry that coalesced around commodity production. On top of this, the supply needs created by World War II, which quickly turned Canada's resurgent agricultural surpluses into shortages, drove up prices for wheat and other cash crops, encouraging greater specialization and boosts in production. World War II unleashed the same trends in other countries, making extensive state intervention, in the form of planning and supply management, the norm in most OECD countries during the postwar period.[12]

Canadian historian Vernon Fowke gives credit for postwar

agricultural policies in Canada to the strength of farmers' organizations at the time. More recent researchers, such as Anthony Winson, acknowledge the critical role played by farmers' organizations and agrarian-based parties in agitating for producer co-operatives and state-led forms of intervention, as a means of controlling the market's volatility and tendency towards monopoly. However, they also argue that it is more accurate to see state policies as the confluence of several forces. For Winson, the most important among these were the federal government's wartime policies and industry's eventual support for state intervention, once it became clear in the 1930s that the entire Western agricultural economy might collapse if left to free-market forces. With this convergence of interests, the politically tense first decades of the twentieth century settled down into a roughly thirty-year postwar period of consensus and stability around agricultural policy.[13]

At the farm level, by contrast, these were not stable times. The farm was changing dramatically. From 1931 to 1961 the number of farm machines in Canada increased from 162,000 to 1,398,000. Fertilizer use surged as well, with expenditures on fertilizers rising from about $10 million in 1931 to over $77 million in 1961. Chemical pesticides were introduced at the end of the 1940s and were widely adopted by the late 1950s. Farmers may have been relatively content with the transformations of their farms wrought by specialization and new, industrial technologies. The high, stable, commodity prices of the postwar period, and the immediate benefits and hidden costs of the new technologies, would have dulled much of the opposition. But there were also strong reasons for discontent. The new technologies set in motion a technological treadmill that between 1931 and 1966 saw the number of farms in Canada drop from 729,000 to 481,000 – with the average acreage of farms nearly doubling from 224 acres to 404 acres.[14]

Agricultural Research and Postwar Production

Although government intervention, largely in the form of public agricultural research and extension, was behind this farm-level transformation, agricultural research is notably absent in both Fowke's and Winson's accounts of postwar state intervention in agriculture. If, as Winson maintains, government intervention in agriculture was primarily "a result of the pressures of organized interests from below," can the same be said for the increasing government intervention into agricultural practice?[15] Agrarian organizations that championed the cause of greater farmer control over the off-farm agricultural system (transport, marketing, banks) were quite silent when it came to the situation on the farm, where specialization, mechanization, chemical inputs, public seed varieties, and new extension programs were rapidly changing the farm and increasing farmer dependence on outside actors and agencies. In fact, the producer co-operatives, for which farmers fought so hard, became the primary vehicles for the promotion and supply of industrial agricultural commodities such as fertilizers and pesticides.[16]

Some writers have highlighted the efforts of small-farm populist movements in the U.S. and Canada to oppose the imposition of industrial agriculture during the postwar period. But this opposition was marginal, and not even influential on the left of the political spectrum. The major actors – industry, government, and producers – and both sides of the political spectrum had coalesced around a utilitarian, productionist paradigm: the surplus, industrialized production of certain agricultural crops for the satisfaction of perceived national interests. During those days, most leftists believed that the industrialization of agriculture was justifiable by the overall benefits generated for the majority, in the form of a cheap supply of food for the working class, while the negative impacts on small-scale farmers could be assuaged through government intervention.[17]

Such a postwar productionist paradigm and its national values were easily integrated into the public agricultural research system, already inclined towards reductionist science and the adoption of industrial technologies. In the absence of any critical challenge, agriculture research became simply a matter of applying the laws and methods of modern Mendelian science to increase profit and production. All other potential indicators and alternatives were deliberately excluded from the framework.[18] And the state, as the supposed "expression of the public interest," became the logical site for agricultural research, with public scientists working for the benefit of farmers, industry, and the Canadian public.

The postwar period, then, also served to centralize and standardize agricultural research. The farmer-to-farmer model of agricultural exchange was quickly replaced by the one-way technology transfer model (scientist-to-farmer), and the diversity of earlier years gave way to increasing uniformity. The impact was dramatic. In the early decades of modern public agricultural research, from 1920 to 1940, the production of wheat, which was at the centre of public research, more than doubled, while the production of other crops, once mainstays of Canadian agriculture, was neglected by agricultural research, dwindled and, in some cases, practically disappeared. The production of buckwheat dropped from 8,965 million bushels in 1920, to 6,692 million bushels in 1940, and then to a mere 2,938 million bushels in 1950. While the diversity within crops like wheat was lost with the uptake of new public varieties, diversity in the crops less favoured by agricultural research simply faded away. The Tartarian variety of buckwheat, for instance, highly prized in Quebec and the Maritimes where its flour was used for pancakes, is now nearly extinct.[19]

Canada's Hybrid Corn Miracle?

As certain crops were neglected and marginalized, their productivity tapered off too. The production of corn, perhaps the most important crop in North America before the arrival of the Euro-

Table 2
Production of Agricultural Crops in Canada
in 1920 and 1940

Crop	Number of Bushels in 1920 ('000)	Number of Bushels in 1940 ('000)
Wheat	263,189	540,000
Barley	63,311	135,160
Oats	530,710	380,526
Buckwheat	8,965	6,692
Dry Peas	3,528	1,355

Source: Statistics Canada, *Handbook of Agricultural Statistics: Field Crops*, part I, (Catalogue 21–516); Canada, Dominion Bureau of Statistics, *Handbook of Agricultural Statistics: Field Crops*, part I, (Catalogue 21–507).

peans, and a major crop on settler farms in the nineteenth century, declined dramatically during the early part of the twentieth century because it was not part of an organized commodity chain. As corn acreage dwindled, so too did its productivity (see Table 3).

Part of the explanation for the declining yields in corn lies in the lack of formal plant breeding work on the crop. Corn, unlike wheat, is a cross-pollinating crop. It does not lend itself to the pure-line method of Mendelian plant breeding increasingly popular among plant breeders at the turn of the century because its penchant for diversifying itself made it difficult to "lock in" particular traits.[20] It was a source of endless frustration for these plant breeders, and, with Mendelian genetics taking over as the dominant school of thought at the beginning of the twentieth century, formal breeding work on corn pretty much ceased by 1910.

As public breeding of open-pollinated corn waned, the development of hybrid corn south of the border rekindled interest in corn breeding among Canadian scientists. To produce hybrid seeds, breeders begin by selecting a number of crop lines with desired characteristics and self-pollinating them. The inbreeding

Table 3
Production of Corn in Canada 1910–74

Year	Total acreage ('000 acres)	Total produc- tion ('000 bushels)	Average yield°
1900	361	25,876	71.68
1908	366	22,868	62.48
1910	294	14,318	48.70
1915	253	14,368	56.79
1920	292	14,335	49.09
1925	239	10,564	44.20
1930	161	5,826	36.19
1935	168	7,765	46.22
1940	186	6,956	37.40
1945	243	10,635	43.76
1950	311	14,103	45.35
1955	571	35,558	62.27
1960	456	26,099	57.23
1965	746	59,498	79.76
1970	1,234	103,684	84.02
1974	1,460	101,910	69.80

° determined by dividing the total production by the total acreage
Source: Statistics Canada, *Handbook of Agricultural Statistics: Field Crops*, part I, (Catalogue 21–516); Canada, Dominion Bureau of Statistics, *Handbook of Agricultural Statistics: Field Crops*, part I, (Catalogue 21–507); Canada, Dominion Bureau of Statistics, *Sixth Census of Canada, 1921: Agriculture*, vol. V; Canada, Department of Agriculture, Census Branch, *Census of Canada, 1880–81*, vol. III; Canada, Department of Agriculture, Census Branch, *Census of Canada, 1870–71*, vol. III, Statistics Canada, *Quarterly Bulletin of Agricultural Statistics*, (Catalogue 21–003); Canada, Department of Trade and Commerce, Census and Statistics Office, *Census and Statistics Monthly*; Canada, Department of Trade and Commerce, Census and Statistics Office, *Fifth Census of Canada, 1911: Agriculture*, vol. IV.

is then repeated with subsequent generations in order to isolate "pure lines" – plants that produce exact clones of themselves when inbred. With these pure-line varieties, the breeder then experiments by crossing one pure line with a different pure line. Some of the crosses should result in a first generation of plants that are completely uniform and, due to a controversial phenomenon called "hybrid vigour," higher-yielding than their inbred parents. But seeds from subsequent generations will produce plants that perform unevenly and poorly, obliging the farmer to purchase new seed every year, thereby increasing seed sales. Hence the immediate interest among seed companies in the development of hybrid varieties.

Some people think that hybrid corn was less a technical success than a political one. French agronomist Jean-Pierre Berlan even dismisses the notion of "hybrid vigour" as a myth and claims that hybrid technology is fundamentally flawed. He notes that the initial inbreeding steps for hybrid seed production generate thousands of pure-line varieties that may all potentially produce excellent hybrids when crossed with another pure line. A measly looking pure line could produce a robust high-yielding hybrid when crossed with another measly looking variety. This leaves no way for a breeder to tell which cross will be most productive, without carrying out all of the potential crosses, something that no breeder can actually do. So, Berlan says, the breeder is inevitably forced to work with a small subset of lines, in effect reducing the genetic base of the original population and, in so doing, making "improvement" impossible. For Berlan, if breeders succeeded in developing hybrid varieties that outperformed the open-pollinated (non-hybrid) varieties, it is only because they did something else: they abandoned alternative breeding methods to focus exclusively on hybrids.[21]

In Canada, although the U.S. hybrids that were introduced first did not perform well, Canadian scientists were convinced of the value of the technique. So much so that in 1923 the Harrow, Ontario, research station, the leading centre for research on corn among the federal stations, decided to focus exclusively on the development of hybrid varieties. It took sixteen years for the station to release its first variety. It then took another fifteen years for the hybrids to match the yields that farmers had been realizing with open-pollinated varieties thirty years earlier, when the public sector had abandoned research on open-pollinated varieties.

Were these hybrid corn varieties a success? Farmers did take up hybrid varieties in a dramatic fashion: by 1944 practically all the corn acreage in Canada was sown to hybrids. But the hybrids were launched as part of a larger agricultural project to increase corn production that brought together government, the seed industry, and the downstream processing industry. Had public breeding programs chosen to pursue the development of open-pollinated varieties instead of hybrid varieties in 1923, they may have produced similar, if not higher, yield increases.

Recent studies looking back at the history of hybrid corn breeding in Ontario indicate that the genetic yield gains of the past fifty years are due mostly to the increased ability of corn plants to tolerate stress. One group of researchers from the University of Guelph concludes: "New hybrids suffer less yield reduction under conditions of drought stress, high plant population, weed interference, low nitrogen, herbicide injury, and low night temperatures. These changes in stress tolerance are likely the by-product of plant breeders selecting for yield at high plant populations and over a wide range of growing environments." In other words, the yield increases are the result of population breeding efforts that have little to do with hybrid vigour or Mendelian breeding techniques.[22]

Other factors may have been at work, but the Mendelian culture pervading the Canadian agricultural research system at the time certainly influenced the choice to pursue hybrid corn. The same goes for other crops, where politics and ideology were just as central as biology in determining the course of plant breeding. When Charles Saunders, who by then had become head plant breeder within the Ministry of Agriculture, resigned from his post in 1921, he was replaced by H.L. Newman, another staunch follower of the Mendelian school. Newman, in a quest to conquer the wheat stem rust then ravaging Canada's wheat crop, took the national wheat breeding program ever deeper into Mendelian genetics. International conferences were organized, a Central Rust Research Laboratory was set up, and, twelve long years later, a rust-resistant variety, Renown, was released. By then, plant breeding in Canada was firmly anchored in centralized Mendelian breeding programs, with the door tightly shut on alternative paths that could have allowed for more diversity and participation.

Building a "Public" Seed System

By the time of World War II, the federal government had become the central actor in the Canadian seed system, intervening directly to establish a national seed system based on public breeding programs. But the state was not the only actor involved. Most of the research and development into industrial agricultural technologies, such as tractors, agricultural implements, and chemical inputs, was undertaken by the private sector. Seeds were an exception. Even as late as the 1980s there was practically no private-sector plant breeding, with the public sector still accounting for over 95 percent of formal plant breeding in Canada and 100 percent of the breeding for cereals and oilseeds.[23]

There were various reasons for this exceptional government intervention. Throughout the first half of the twentieth century, the private seed sector consisted of only small, private companies. Few were engaged in plant breeding; they focused more on the multiplication and distribution of varieties brought in from the U.S. and other countries. There simply was not much interest in plant breeding within the private sector. Canadian seed markets were relatively small and difficult to breed for because of the country's unique and diverse agricultural and climatic conditions.[24] Moreover, the state's agricultural development policies focused on standardized export commodities, making it very difficult for private breeding programs to develop and introduce new, "improved" varieties that could meet the high standards set by the *Canada Seeds Act* of 1923. This required all new varieties to pass through tests proving superior performance over existing ones. Without a private sector to turn to, the state felt it had to step in and take over the bulk of formal research and development in plant breeding, limiting the private sector to "final stage" activities – marketing public varieties, or developing new varieties using public varieties.

From a broad, social standpoint, the investment in plant breeding was easily justifiable because the returns on investment were measured by the overall welfare gains they brought, not the seed sales they generated. Public plant breeding was a matter of national interest, undertaken to benefit farmers, consumers, and industry, and there was little opposition to this perspective. In the words of one official from Agriculture Canada, speaking in 1971, "A strong Canadian seed system is worth promoting, especially in this period of nationalism."[25]

This spirit of working in the national interest pervaded the public plant breeding system and the various institutions, public and private, in Canada and abroad, that collaborated with it. In his study of the development of canola in Canada, food systems analyst Brewster Kneen says that Canadian researchers and their intended beneficiaries possessed a "common culture." According to Kneen:

> Everyone involved [in the development of canola] shared a common culture and concern. . . . They were all white males of European/British descent; and, they virtually all grew up on Prairie farms during the Depression and shared a profound and unique experience that their followers, born and raised elsewhere, whether in cities or eastern farming areas, could never know. . . . Neither in conversation nor in the available literature does the issue of the appropriateness of such elitist decision-making or of accountability ever come up. It would appear that the whole process was – and

still is – regarded as a technical matter to be decided by the individuals involved directly or by committees of experts in the field. There is no hint of farmers, except as some of them were represented on the Canola Council, or people involved in other agricultural issues, such as soil conservation or long-term sustainability, ever being consulted. Of course, farmers were much more numerous and constituted a significantly higher percentage of the Canadian population at mid-century than they do now, and the culture of the country was itself much more cognizant of its rural and agricultural components, or simply more agrarian. So it was not entirely unreasonable for the researchers to assume that they were acting in the best interests of the farmers and the country."[26]

With canola (otherwise known as rapeseed), as with other crops, seed "improvement" was initially carried out by farmers. But once the public sector and the state moved in, farmers were marginalized from plant breeding and the farmer-to-farmer seed exchanges were replaced by top-down "technology transfers" from scientists to farmers. We can partially attribute the lack of farmer resistance to this transformation to the "common culture" of farmers and researchers, as well as to the overriding productionist coalition that melded farmer, industry, and state interests. But there are other important reasons that should not be overlooked.

While it is true that with the emergence of the public plant breeding programs farmers lost their central place in plant breeding, they continued to play an important role in the larger seed system by multiplying, distributing, and saving seeds. Farmer participation in the seed system was institutionalized at the very beginning of the public plant breeding programs with the formation of the Canadian Seed Growers' Association (CSGA), an association of farmers. When public programs released varieties, the breeders sent certain CSGA members seeds to carry out the first two generations of multiplication. The seed was then distributed to more CSGA members who multiplied it into registered, and then certified, seed. The certified seed was then sold to farmers. From this point on, farmers continued to take care of the seed by saving it for themselves or their neighbours for any number of generations, depending on the crop. The few studies that exist indicate that farmers were quite capable of maintaining seed quality through seed-saving. According to a study in Alberta in 1980, 60 percent of the farmer-saved seed surveyed was equal to the highest quality seed on the market.[27] The reality is that farm-saved seed never stopped providing the bulk of Canada's seed supply. At the end of the 1970s, only 20–30 percent of seed used in Canadian agriculture was certified seed; there was only enough certified seed available for 14 percent of the seeded acreage for wheat, 31 percent for barley, and 30 percent for oats.[28]

Farmers could also take some comfort in the knowledge that their interests were protected in the legislative framework built into the national seed system. The *Seeds Act* of 1923 was designed to prevent sales of poor quality seed to unsuspecting farmers and to prevent seed companies from making false claims about the performance of their varieties. It restricted sales of seed to registered varieties and set a high standard for variety registration: under a 1929 amendment new cereal varieties could only be registered if they proved to be superior to the best variety already on the market. Moreover, the main institutions of this registration system were the Registration Recommending Committees in which farmer representatives participated. These committees had a large influence over what varieties were registered as seed and a less direct, but important, influence over plant breeding.[29] The downside of the *Seeds Act* was that it made registration compulsory and set standards that few farmers' varieties could achieve, even if farmers had the desire to go through the hoops of registration. This greatly contributed to the decline in crop diversity.[30] Plus, the testing system itself, which centred around the experimental farms, elevated the position of these central stations in the

overall system, and handed to scientists the work, formerly undertaken by farmers, of evaluating varieties.

In conversations over the years with farmers who remember
the early days of the public breeding programs, most have told me
that one of the main reasons why they liked the public programs
was that, quite simply, they provided good varieties. While some
of the statistics may exaggerate the influence of genetics, there is
no doubt that the public plant breeding programs made a significant contribution to increasing crop yields.[31] And beyond yields,
the public breeding efforts achieved important successes in
breeding for disease resistance and in developing new crops identified by farmers, industry, or the state, such as hybrid corn, short-
season soybeans, and canola. Moreover, and this is a big plus that
farmers today are keenly aware of, the costs of plant breeding
were not passed solely on to farmers through the price of the
seed; they were shared by society as a whole through public
expenditures. The prevailing assumption of the time was that the
development of food and agriculture served the entire society –
farmers, consumers, business – and, accordingly, the costs of public plant breeding should be shared. And with seeds as public
goods there were no pressures, such as exist today, for farmers to
buy seeds or pay royalties every season. Seed-saving was indeed
the most direct way to multiply the value of the seeds released by
the public programs.

The collective nature of the public breeding programs was
reflected in its more practical plant breeding aspects too. Successful public breeding efforts could never be the work of particular
breeding programs acting alone. W.T. Bradnock, the Director of
the Seeds Division of Agriculture Canada during the latter part of
the 1980s, put it this way: "Plant breeding resembles the creation
of a mosaic with contributions from different sources needed to
complete the design."[32] Bradnock was referring to collaboration
and the exchange of breeding material, known as germplasm, with
other plant breeding programs, but his metaphor could also have

encompassed farmers, since nearly all of the varieties used in public breeding programs (and private as well) were originally developed by farmers, occasionally from Canada but most often from other parts of the world. Plant breeding, as it was practised at the time, was a continuous process of collective innovation, flowing in and out of experimental stations and farmers' fields.

The Case of Canola

The development of canola exemplified this "public" model of the second seed regime. Prior to World War II canola, then known as rapeseed, was a minor crop in Canada, produced and looked after by a small number of farmers who had brought varieties over from Europe. But experiences of wartime edible oil supply shortages and concerns about the Prairies' over-dependence on wheat sparked government and industry interest in developing a reliable, domestic source of edible oil that could be grown in the West.[33]

Public and private work on the crop began during the war and intensified two decades later in the 1960s. In 1967 the federal government set up the Rapeseed Utilization Assistance Program and the Rapeseed Association, which brought together all the relevant actors to help guide federal research and policy: scientists, growers, handlers, crushers, exporters, refiners, and feed manufacturers. All of the initial participants were Canadian, and the priorities, to establish a domestic edible oil supply, build export markets, and help diversify prairie agriculture, were based on perceived national interests. This common vision fostered collaboration and openness. According to Brewster Kneen,

> The overall milieu was one of a self-selected group of government, industry and university men of highly similar background working together at a common task, largely unconcerned as to who got the credit, and, apparently, largely without institutional chauvinism.

Their institutions were regarded as the tools or facilitators of the research, not the proprietors.[34]

Knowledge and breeding material were freely exchanged among canola researchers, since everyone understood that they were working collectively in the same direction. Those involved in the development of canola credit the success of the project to this collaborative and collective culture. According to one 1975 article describing the development of canola:

> Cooperation – this has been the most important aspect of the rapeseed story. Though emphasis has been placed on the teamwork among the scientists, it existed throughout the rapeseed industry as a whole: among farmers, oilseed processors, and businessmen of the food industry. The exchange of information in the arena of international agricultural science was also important. Without this cooperation, devoid as it was of formal structuring, rapeseed might have remained for Canadians what it was in the early stages of development – a laboratory curiosity.[35]

Kneen, too, credits the successful development of canola to the "informal but highly functional research culture" of co-operation among federal researchers, universities, and industry.[36]

The results of this collaborative effort are not so positive if we look at the larger picture. Canola, originally meant to diversify prairie agriculture, has become another intensive monoculture crop, useful for rotations, but rarely grown as part of an integrated agricultural system. In this, its evolution resembles that of the other commmodity crops developed during this period. But the productionist coalition that dominated agricultural research at the time, and the unquestioned assumptions around "public interest" or "national objectives," facilitated the functioning of the seed system by giving plant breeders a significant amount of leeway in their research and a free and open environment in which to work.

It is not the collaborative culture or the collective spirit of the public seed system that are problematic, then, but the lack of public oversight and broad-based engagement in the research process. As Kneen points out,

> The high degree of cultural homogeneity was, according to the records, not tempered in any way by what might be referred to as a larger democratic process. . . . This is not to say that the decision was wrong or contrary to what might have been decided if some larger democratic process had been operative. But when we come to look at the research agenda today and how it is determined and by whom, to say nothing of who pays for it, we might well wish that long ago there had been instituted some means of ensuring public debate on the subject.[37]

These days, at the beginning of the twentieth-first century, the public seed system is on its last legs. The open culture it thrived in is disintegrating along with it. Many of us opposed to the corporate seed system that is taking its place may welcome the decline of the productionist coalition and its institutions. But we are likely to find the roads leading in alternative directions quite arduous. The culture and values that support the status quo are deeply ingrained in the Canadian agricultural landscape, even though, as we shall see, they have been corrupted to serve the narrow interests of the corporate seed industry.

Chapter 5

Corporation

In the 1970s and 1980s the productionist coalition that characterized the post-World War II seed system in Canada began to disintegrate. The references to national interests that lay behind its broad ideological support had started to lose their hold. With increasing public awareness of the environmental, social, and health costs of productionist agriculture, and with agriculture's economic weight relative to other sectors in decline, much of the public was no longer willing to absorb these "externalized" costs. Producers were suddenly confronted with new political actors, from rural property owners to environmental groups, who were willing and able to engage in local and provincial policy struggles over agricultural practice.[1] People were also reacting against the "bulk commodity" approach to food supply, and by the 1990s this was giving rise to sizeable niche markets for organic and "high-quality" foods.

Yet there was, and still is, more than simple consumer dissatisfaction involved here. Producers may regularly condemn the supposedly cheap price of food for keeping farm-gate prices down, but the lineups at urban food banks continue to grow. The poor are increasingly done a disservice by the current agriculture and food systems, in terms of both access to food and food quality.[2] If social activists and progressive intellectuals originally backed productionist agriculture for its potential to bring benefits to farmers and the poor, by the 1980s few could deny that the main beneficiary of the system was agribusiness. The clear losers were small farmers and the poor. Many, therefore, began to desert the productionist coalition and turn their attention to a "new agrarian question," calling for an alternative direction in food and agricultural policy that would marry changes at the farm level to larger transformations of the food system based on equity.[3]

Meanwhile, agribusiness was also departing from the productionist model. While the ideological shift among social activists and progressive intellectuals was based on a critique of postwar productionism, international developments in the food and agricultural industry were already setting the stage for the old model's decline. For two and a half decades, while other sectors were subjected to deep, global restructuring, agriculture and food endured within national regulatory systems held together by complex and often bizarre webs of domestic and international political ambitions and alliances. But after the food crisis of the 1970s, in which global food surpluses rapidly turned into shortages, the old order began to fall apart.[4]

The new food order that emerged was shaped by a vast, corporate restructuring in global agribusiness, driven by consolidation in the retail sector.[5] In Canada, by 1987 the five largest grocery distributors accounted for 70 percent of all grocery sales, with concentration being even more significant if cities or regions are considered separately. Retail concentration translated into great economic power over other sectors in the agri-food industry, particularly producers. From 1981 to 1987, as farm gate prices decreased by 10 percent, retail prices rose by 32 percent. Other sectors of the agri-food industry were also affected and forced to respond.[6]

Pesticide corporations, squeezed by declining producer revenues and generic competition, reacted by buying up seed companies and investing heavily in biotechnology. They realized early on that they could strengthen their position vis-à-vis the rest of agribusiness through the proprietary control of seeds and genetics (the primary elements in the agri-food chain), which they could

transform with "value-added" properties.[7] The large food processors and commodity traders, such as Cargill and Archer Daniels
Midland (ADM), were also moving towards "product differentiation" or "value differentiation" based on proprietary technology.
They saw this as a means to escape from the downward pressure
on prices coming from their buyers, who were in a position to play
them and their competing bulk suppliers off against each other.[8]
The newly refashioned Cargill Company tellingly refers to its
work as "food systems design."[9]

Overall, this corporate positioning has intensified vertical integration all along the agri-food chain. Through mergers and
alliances, firms are able to exercise greater control over their suppliers and gain greater leverage against their buyers by, in effect,
substituting internal organization and transactions for the market.
Industry analysts maintain that the current trend is towards tightly
integrated "supply chains" or "food systems" organized around a
single firm or alliance of firms and operating from "field-to-fork,"
with some economists speculating that farmers will soon be identifying themselves as members of the particular corporate "food
systems" of which they are a part.[10] This, however, does not mean
the end of bulk commodity production. Even as integrated supply
chains evolve, most agricultural production destined for the market will continue to be of the traditional commodity crops, such as
cereals and oilseeds. The grain merchants do not want to relin-
quish their economies of scale just to satisfy some niche market
opportunities for so-called value-added grains that they have to
segregate carefully. The difference, however, will be that this production will not occur within the national regulatory frameworks
established during the postwar period, which provided at least some
protection to farmers and other domestic interests. In the new
global food system, farmers and the crops they produce will be, in
the business lingo, more and more "substitutable" – easily replaced
when a lower-cost source appears. Freed from government-
imposed regulatory restraints, transnational corporations will be in

an increasingly powerful position, setting the game according to their own rules and regulations.[11]

Canada, like most other OECD countries, has generally not resisted this restructuring. The Canadian government's response has been to move gradually away from the postwar national policies and push for a more liberalized trading environment, with foreign investment as the engine of economic growth. The interventionist state of the postwar period, which focused on broad national economic objectives and was sensitive to the interests of various constituencies, has largely given way to a neo-liberal state concerned with establishing a business environment conducive to the corporate agenda.[12] So when in the 1980s the corporate sector in the U.S. started investing heavily in recombinant DNA research (genetic engineering), the Canadian government launched another round of policy initiatives in the form of a co-ordinated national strategy to develop a national biotechnology sector that could attract this latest hot target for foreign investment.

The National Biotechnology Strategy

The corporate sector had a clear grasp of the potential commercial applications of recombinant DNA technology (genetic engineering) by the end of the 1970s.[13] In 1980 an executive with DuPont proclaimed: "The [biotechnology] bus is moving . . . and if you want a ticket you'd better get it now."[14] At the time, nearly all commercial biotechnology activity took place in the United States, but by the beginning of the 1980s the message had crossed the border, carried by Canadian scientists from government and universities eager to keep up with their counterparts to the south. These scientists argued that Canada would miss out if it did not act quickly and they took their message directly to the federal government.

The government responded in 1982 with the creation of a Task Force on Biotechnology which would, one year later, produce the

first National Biotechnology Strategy. According to Lewis Slotin, then a policy advisor to the Ministry of State for Science and Technology, and one of the "architects" of the National Biotechnology Strategy, the process to develop the strategy began "with one simple objective: and that was to create an awareness in the country that biotechnology was going to be damn important for the future of our competitive position."[15]

The Task Force was a mix of men from academe and industry who saw in the science of biotechnology an industrial vision. As in other OECD countries, Canadian biotechnology policy was, from the outset, an industrial strategy to ensure Canada's economic competitiveness. According to the Task Force members:

> Throughout the deliberations the Task Force was well aware of the advantages of a "market-pull" rather than a "technology-push" approach to industrial development. However, the almost total absence of biotechnology industrial activity in Canada necessitated recommendations supporting a technology orientation, at least in the short term, for this country's development of biotechnology.[16]

In March 1987 federal biotechnology efforts were integrated into a more systematic National Science and Technology Policy. Through it the government created a National Advisory Board on Science and Technology (NABST) and merged the Ministry of State for Science and Technology and parts of the Department of Regional Industrial Expansion to form a Department of Industry, Science and Technology. Then it launched InnovAction – described as a five-track strategy to lead Canada's science and technology efforts, under which biotechnology was identified as one of three areas of strategic technology "paramount to Canada's international competitive position." The launch of InnovAction was followed up a year later with a major new funding commitment to science and technology of $1.3 billion over five years, privileging the three strategic sectors.[17]

It was a significant change in federal policy. The new approach was outlined by Prime Minister Brian Mulroney in a 1989 address to the NABST:

> The goal is an economy that can compete with the best in the world, producing stimulating new jobs and new opportunities for future generations of Canadians. . . . Science and technology are the keys to a modern competitive economy. It is clear that our traditional manufacturing and resource-based industries will no longer assure us a strong position in the global economy if we don't complement them with modern technology.[18]

Under the Mulroney government, even as federal spending on research declined, science and technology moved to the centre of Canada's industrial/economic policy – not because the government was giving more importance to science, but because it began to conceive of science and industry as one and the same.

Building up the biotechnology industry, however, would take more than state funding and a short-term "technology orientation." Despite strong public financial support, by 1993, a decade after the launch of the National Biotechnology strategy, the industry had little to show for itself; there were only thirty firms involved in agricultural biotechnology – nine of these were public institutions (university or government) and seven were foreign TNCs, such as Monsanto and DuPont. None of the fourteen private Canadian firms involved in agricultural biotechnology, such as Allelix or Paladin Hybrids, was profitable.[19] As one of the leading figures in the development of Canada's biotechnology industry cautiously pointed out: "In Canada there is the danger that governments intervene to maintain non-competitiveness of companies perhaps beyond their natural stage of things. . . . I'm not sure just how many of Canada's [biotech firms] would survive on their own."[20] If the state wanted to build up a biotechnology industry, it would have to get involved in a more profound way, by building up an entire regulatory structure.

Canadian biotechnology regulations emerged in the context of a larger policy shift at the federal level. In the mid-1980s the federal government introduced two major statements that laid out a new direction for regulatory policy based on two specific objectives: to reduce impediments to economic growth, and to remove obstacles to innovation. As part of this policy, the government moved overall regulatory responsibilities to the Privy Council Office and the Treasury Board, where the regulatory policies of the different departments could be heavily policed.[21]

With the overarching structural reforms in place, the government then turned to the policy arena. In the Budget Papers of February 1992, under the heading "Tackling the Regulatory Burden," the minister of finance announced:

> The government is beginning a department review of existing regulations to ensure that the use of the government's regulatory powers results in the greatest prosperity for Canadians. In this context, departments will be instructed to review their existing regulations to ascertain whether they comply with this objective. This is a major undertaking and will have to proceed in stages. Agriculture Canada, Transport Canada and Consumer and Corporate Affairs will be the first departments to engage in this review. Part of this review should require a public "rejustification" of existing regulations that are to be retained to ensure that those which stifle the creativity and efficiency required by Canadian business to compete and grow in today's modern world or which serve no public good, are removed.[22]

Three days later, the newly established Regulatory Affairs Directorate in the Treasury Board published an update on the federal government's regulatory policy. According to this update, departments had to demonstrate, among other factors, that for existing or proposed regulations:

- a problem or risk exists, government intervention is justified, and regulation is the best alternative;

- the benefits of the regulatory activity outweigh the costs; and,
- steps have been taken to ensure that the regulatory activity impedes as little as possible Canada's competitiveness.

These policies had the immediate effect of making it much more difficult for departments to enact new regulations or even maintain existing ones, particularly if they clashed with industry interests.

In 1992 the cabinet made an official decision to develop a federal regulatory framework for biotechnology, even though biotechnology policy discussions had taken place behind closed doors for some time. Through close consultation with industry and regulators in the U.S., the delegated Subgroup on Safety and Regulations came back with a proposal that has since come to define Canada's regulatory framework for genetically modified organisms. They recommended a broad definition of biotechnology that was subsequently adopted in the 1993 *Canadian Environmental Protection Act*: "the application of science and engineering in the direct or indirect use of living organisms or parts or products of living organisms in their natural or modified forms." The Subgroup also insisted that genetically modified organisms be considered on a product, not on a process basis, leading to the division of responsibility among a number of agencies. The key to the proposal was the notion of "substantial equivalence," developed within the Organization for Economic Cooperation and Development (OECD) with the active participation of the Subgroup and colleagues at Agriculture Canada. According to Simon Barber, who arrived at Agriculture Canada's Food Inspection Branch in 1990 to set up the regulatory framework: "We propose that . . . if the new product is 'substantially equivalent' to existing, familiar products accepted as safe . . . we could then waive further requirements for risk assessment."[23]

The Subgroup's recommendations, which were eventually adopted, were entirely in step with the directions coming down from the Treasury Board and the central bodies. With some clever use of language, they brought a new and unknown technology

under existing regulations without creating new regulatory "burdens" for the industry.

When it came to safety issues for their GM crops, the biotechnology and seeds corporations wanted minimal government regulation, and this is what they got. Regulation here, they argued, would throw up a barrier to investment. But there are other areas of regulation where these corporations have pushed for ever-greater controls to protect their investments. The biotechnology lobby, for instance, was instrumental in bringing in stronger patent protection for pharmaceuticals during the final days of the Conservative Mulroney government. With plants too, the biotechnology lobby succeeded in bringing in legislation giving seed companies patent-like rights over plant varieties, and creating enormous regulatory burdens for everyone else in the seed system.

It is often said that corporations want deregulation. But this is not always the case. Where regulations impinge on corporate profits then, yes, corporations want deregulation; but where people's practices impinge on corporate profits, corporations want regulations or other forms of government intervention to protect their interests. In the case of seeds, as we have seen, the main impediments to corporate profits were the high levels of seed-saving among Canadian farmers and the open and collective public breeding programs. Public breeders supplied decent varieties at good prices to farmers who were then free to do what they wished with them. Seeds, in short, were more public goods than they were commodities. If there was going to be more corporate investment in the seed industry, if there was any hope of shifting the centre of power in the seed system from the public breeding programs to corporate boardrooms, then this picture had to change – seeds had to be commodified.

Chapter 6

Commodification

THE MOST STRAIGHTFORWARD WAY to stop farmers from saving seeds, without having to resort to complex and difficult-to-enforce regulations, is to develop plants to produce seeds that are either sterile, low yielding, or unable to grow properly without the use of certain chemicals. The hybrid breeding method did this for corn early on in the twentieth century, and private seed companies moved in quickly to take control of the corn seed industry. Seed companies also had success developing hybrids for some vegetable and some cereal and oilseed crops, like sorghum. And with these too, private sector plant breeding soon displaced public. But with many of the major cereal and oilseed crops, like wheat and soybeans, technical difficulties with the breeding process continue to make hybrid seed production uneconomic and impractical.

More recently, genetic engineering has opened up a new biological route for the seed industry to commodify the seed, in the form of genetic use restriction technologies (GURTS). Commonly known as Traitor or Terminator technologies, GURTS are genetic engineering techniques that modify plants so that the seeds they produce will not germinate if planted. The technique involves a method whereby a gene necessary to plant growth can be turned on or off in a developmentally regulated fashion, as well as a procedure for controlling from the outside the expression of an engineered gene, using a chemical inducer or other factor, such as cold treatment. A farmer saving seeds would conceivably have to spray these seeds with a particular chemical in order for them to germinate.

While GURTS technologies have yet to be commercialized, as of late 2005 there were twenty-two different patents on GURTS granted or applied for under the *Patent Cooperation Treaty* of the World Intellectual Property Organization. In October 2006 the Canadian Intellectual Property Office granted a controversial GURTS patent to Delta & Pine Land, a U.S. cotton seed company now owned by Monsanto, that claims to have advanced a GURTS technology to the point of laboratory and greenhouse tests with tobacco and cotton plants.[1] At this point, however, the Terminator technologies are still in the pipeline and it remains unclear if, and when, they will come to market. In Canada, where there has been a strong campaign to ban the technology, a bill to prohibit field testing and the commercialization of Terminator technologies was introduced in Parliament on May 31, 2007.

So, in the absence of a feasible and near-term biological option to stop seed-saving, government and industry have turned instead to a whole array of other mechanisms, from patents to contracts, to police the countryside and make seed-saving, to all intents and purposes, a crime.

Step One: Patents

Patents are legal instruments that provide monopoly rights over a creation for commercial purposes over a period of time. A patent is a right granted to an inventor to prevent all others from making, using, and/or selling the patented invention for fifteen to twenty years. The criteria for a patent are novelty, inventiveness (non-obviousness), utility, and reproducibility. Although patents were designed for industrial application, in the realm of biotechnology patent offices now grant patents on genes or genetic constructs and, in some countries such as the U.S., on plant varieties.

Canada, however, does not allow patents on seeds or plant varieties. In 1987 the Canadian Intellectual Property Office (CIPO) rejected a patent application by Pioneer Hi-bred for a soybean variety – the first application of its kind to come before the CIPO. Ever since, the CIPO has consistently refused patents on higher life forms, a policy that was supported in 2002 by the Supreme Court in its ruling in the closely watched Oncomouse case.[2] However, back in 1982, the CIPO, in applications by Abitibi for a yeast culture and Connaught Laboratories for a cell line, had recognized patents on unicellular life forms and gene sequences. In other words, you could patent genes in plants but you could not patent the plant itself. So what would happen if a company were to genetically modify a plant using a patented gene or gene construct? Would the patent on the gene give the company rights over plants containing that gene?

The full implications of the 1982 decisions only became clear a couple decades later. Percy Schmeiser, a farmer in Bruno, Saskatchewan, had grown canola since the 1950s. The last time he claims to have purchased seed was in 1993. Over the years he says he has saved seed and, through selection, has been able to develop his own strain relatively resistant to various diseases. In 1996, though, Monsanto introduced into Saskatchewan its Roundup Ready canola, a plant genetically engineered with a patented genetic construct to enable it to withstand large volumes of Monsanto's herbicide glyphosate. Two years later the company's private inspectors were in Schmeiser's fields taking samples.

Tests of these samples conducted by Monsanto showed that canola in Schmeiser's fields was glyphosate-resistant and the company immediately took him to court for patent infringement. Monsanto argued that its patent rights on the gene construct extended to all plants that contain it, including the canola growing in Schmeiser's fields. Schmeiser argued that he did not deliberately sow his fields with Roundup Ready canola and that, if his fields were Roundup Ready, it must have occurred by way of an accidental roadside spill of Roundup Ready seed or contamination from cross-pollination with neighbouring fields.

Judge MacKay of the Federal Court ruled that Schmeiser was guilty of a) knowingly having Monsanto genes on his land, and b) not advising Monsanto to come and fetch them. Allegations of obtaining the seed fraudulently were dropped at the hearing due to lack of evidence. It did not matter whether or not Schmeiser was responsible for the Roundup Ready plants being in his fields. Nor did it matter that Schmeiser did not benefit in any way from the Roundup Ready seed. Schmeiser was found guilty nonetheless, and fined nearly $150,000. He also lost the improved genetics resulting from his lifelong practice of saving his own seed – his crop was confiscated.

According to Judge MacKay:

> The defendants grew canola in 1998 in nine fields, from seed saved from their 1997 crop, which seed Mr. Schmeiser knew or can be taken to have known was Roundup tolerant. That seed was grown and ultimately the crop was harvested and sold. In my opinion, whether or not that crop was sprayed with Roundup during its growing period is not important. *Growth of the seed, reproducing the patented gene and cell, and sale of the harvested crop constitutes taking the essence of the plaintiffs' invention, using it, without permission. In so doing the defendants infringed upon the patent interests of the plaintiffs* (emphasis added).

Judge MacKay's decision essentially put the onus on farmers to identify the presence of Monsanto's Roundup Ready genes in their crops and, if they found them, to take steps to remove the plant or seek permission from Monsanto.[3]

Schmeiser later appealed the decision to the Supreme Court, which agreed to re-examine the question of the validity of the patent and, if the patent was indeed valid, to determine whether Schmeiser had infringed it. On May 21, 2004, in a tight, 5–4 decision, the Supreme Court upheld the ruling of the Federal Court. Monsanto's patent was deemed valid. As the majority saw it, a patent can be infringed by the use of a plant or seed into which the patented gene has been inserted. Furthermore, as noted by one of the lawyers intervening in the case, the mere possession of such seeds or plants raises a rebuttable presumption of use. The Court did not offer any guidance as to what would suffice for this presumption to be rebutted. The Court ruled that Schmeiser had infringed the patent even though he never used Roundup (glyphosate) on his canola.

The Supreme Court's decision focused on a narrow interpretation of case law. It treated plants like any other manufactured good that the patent system was designed for, even though the judges repeatedly pointed out the special nature of biological "inventions." They reasoned that the owner of a patented gene should have monopoly rights over any living organism containing that gene, no matter how it got there and even if, as the majority of judges confirmed, the plants themselves were not patentable. It didn't matter that Schmeiser never deliberately planted Monsanto seeds in his fields. He violated the patent merely by growing plants containing Monsanto's patented gene.

The repercussions of this ruling must be considered in light of the current, extensive contamination from GM plants in Canada. Percy Schmeiser is definitely not the only farmer with fields contaminated with the Roundup Ready gene. Since 2002, Roundup Ready canola has been grown on about half the area planted to canola in Canada. In 2002, researchers at the University of Manitoba conducted a survey of twenty-seven certified seedlots of canola. Fourteen had contamination levels above 0.25 percent and three seedlots had glyphosate resistance contamination levels in excess of 2.0 percent.[4] If the certified seed lots are contaminated, it can safely be assumed that almost every canola field in Canada has some plants with the Roundup Ready gene, whether the fields are planted with Roundup Ready canola or not. Even if

Monsanto cannot possibly pursue court cases against every farmer growing crops contaminated with the Roundup Ready gene, it is the precedent of the Schmeiser case that matters. After seeing what happened when Schmeiser went up against the large transnational firm, farmers may think twice about the legal risks of growing non-Roundup Ready canola and may simply opt for the safer route of buying Monsanto's seeds and signing the contract.

The interpretation of the Canadian courts opens up a whole range of ways, some of which they are already pursuing in other countries, for companies like Monsanto to keep farmers from saving seeds. In Brazil, Monsanto tests crops as they are dropped off at the elevators and forces farmers to pay royalties if their crops test positive for the Roundup Ready gene. When it comes to Argentina, where Monsanto has not had all its demands for patent protection met, the company worked with customs officials in European countries, where its patents are recognized, to test exports of soybeans from Argentina. Once these tests showed the presence of its patented genes, Monsanto launched several cases in European courts to force the shippers to pay royalties. This has put immense pressure on the government of Argentina to step in and stop farmers from saving seeds.[5]

Step Two: Contracts

Monsanto rarely relies on the courts to stop farmers from saving seed from its patented plant varieties. Instead, its primary vehicle is the contract. When the company launched its Roundup Ready seeds in North America, farmers wanting to purchase the seeds had to attend a Grower Enrolment Meeting where Monsanto would explain the technology and the rules governing its use. They then had to sign a Technology Use Agreement. Under the terms of this contract, which all farmers still have to sign in order to purchase Roundup Ready seeds, farmers can only use the seed for planting one crop, and the crop can only be sold for consumption to a commercial purchaser authorized by Monsanto. Monsanto dictates what the farmer can do with the seed from the crop and to whom the farmer can sell the crop. Monsanto can also use the contract to control what herbicides the farmer can spray on the crop and reserves the right to make unannounced inspections of the farmer's fields. In the U.S., where the company has a team of seventy-five employees and an annual budget of U.S. $75 million to enforce its contracts, by March 2003 Monsanto had filed seventy-three cases against farmers for not obeying its terms.[6]

The widespread adoption of Monsanto's GM crops, and the use of contracts, have had an immediate impact on seed-saving. According to one study of U.S. soybean production, while the decline in the rate of soybean seed-saving before GM soybeans were introduced was approximately 1.4 percent per year, with the introduction of Monsanto's Roundup Ready soybean in 1996 the rate of decline in soybean seed-saving increased to 2.3 percent per year, essentially doubling in just six years.[7]

Monsanto is not the only company pursuing such contracts. BASF, a German transnational, has developed what it calls the "Clearfield Production System," integrating non-GM herbicide-tolerant plant varieties with a system of herbicide application. BASF Canada is introducing the Clearfield system for wheat, canola, and corn. The company's website proclaims that all of its seed varieties will be sold only under contract as part of the Clearfield Production System.

The Clearfield Production System requires farmers to sign the "Clearfield Commitment." Like the Monsanto agreement, this contract stipulates that farmers can only use the seed for planting one crop, cannot supply the seed to other growers or users, and that all seed produced under the agreement has to be disposed of as commercial grain and cannot be used for planting a subsequent crop. According to BASF Canada: "A grower who has not complied with the Clearfield Commitment will be responsible for administrative charges of up to $100 per acre."[8]

BASF uses the fact that its Clearfield varieties are not genetically engineered as a selling point. But if they are not genetically engineered, there are no recognized patent claims either. And even if the varieties were protected by plant breeders' rights, these rights would not support the scope of the contract. Nothing in intellectual property law in Canada provides BASF or other seed companies with such far-reaching rights. University of McGill legal scholar Richard Gold maintains that this does not matter. The fact that the Clearfield Commitment relates to intellectual property is irrelevant: "A contract can say anything and all of its provisions are enforceable."[9]

More and more seed companies in Canada are selling their seed varieties exclusively through contracts with farmers. For example, C&M Seeds operates an "Identity Preserved Program" in Ontario. It sells several "high-value" varieties of wheat. To purchase seeds from these varieties, farmers have to sign an "Identity Preserved Growers Agreement," which states that the grower agrees "to use only certified seed from C&M"; "not to sell, give, transfer or otherwise dispose of any Identity Preserved Wheat seed to any one for any purpose"; and "not to retain seed produced from IP Wheat seed for the purpose of re-planting or for sale, transfer or other disposition to anyone."[10] Put simply, the farmer has no rights to their own harvest. In effect, a farmer cannot actually purchase the seed, only rent it for a season from its legal "owners."

The expansion of such contracts weighs heavily on the notion of choice for farmers, particularly as these contracts are used by the downstream industry as part of their supply chains. Cargill and Dow AgroSciences, for instance, have developed a specialty canola that Cargill sells under contract-growing agreements with farmers. Only Roundup Ready varieties of this canola are available, and farmers have to sign a Monsanto Technology Use Agreement, pay the technology fees, and cover some of the costs of "identity-preservation."[11]

Step Three: Plant Breeders' Rights

Historically, seed companies have looked to patents as the optimum means of protecting monopoly rights over their plant varieties. But because Canada and most other countries have resisted recognizing patents on plants, seed companies have instead taken a back door route and have lobbied for legal monopoly rights regimes, known as plant breeders' rights (PBR) or plant variety protection, that apply specifically to plants.[12]

In 1961, after four years of negotiations, the Western European governments established the International Union for the Protection of New Plant Varieties (UPOV), the first convention establishing minimum standards for PBR legislation in member countries. The UPOV Convention laid out criteria for granting rights different than patents; breeders had to show that their varieties were novel, distinct, uniform, and stable. But if PBR provided breeders with access to criteria suited to their plants, the trade-off was that they offered narrower rights than patents. With PBR, breeders got rights over the commercial propagation of their protected varieties, but PBR did not restrict farmers from saving seeds, or breeders from doing further breeding with their protected varieties.

UPOV first came into force in Europe in 1968, and shortly thereafter the U.S. signed into being a similar Plant Variety Protection Act. The path to PBR was slower and more difficult for the seed industry in Canada. After years of lobbying, the seed industry finally succeeded in getting the federal government to introduce a bill to establish a *PBR Act* in 1978, but it died on the floor of Parliament. It took another ten years before the government, with heavy support from seed companies and biotechnology firms, reintroduced the bill, which was finally adopted in 1990. Canada's *PBR Act* is based on the 1978 UPOV Convention; it only covers the unauthorized commercial propagation of protected plant varieties, leaving farmer seed saving, and further breeding with protected varieties, outside of its scope.

The *PBR Act* has important consequences for seed-saving. It creates the legal basis for companies to challenge farmers for customary practices, like saving and exchanging seeds, that many still believe to be firmly within their rights. So far, however, seed companies, which are responsible for the enforcement of the *Act* on the ground, have had a tough time changing farmer practices. Through the use of private investigators, seed companies had, by May 1997, reached twenty-four out-of-court settlements with farmers, worth over $240,000. For the general manger of Cargill Seeds, Bruce Howison, "There [was] still a low level of awareness and understanding at the farm level as to what plant breeders' rights are all about and the ramifications of violating them. It is not something farmers are used to dealing with."[13] Realizing that they would have to do something on a bigger scale, a number of seed companies came together later that year to form the Canadian Plant Technology Agency (CPTA), to police and promote the *Act*. But by 2001, the industry had only pursued between forty to fifty cases of infringement worth around half a million dollars in fines.[14]

One plant breeder with a Saskatoon subsidiary of the Swedish multinational seed company Svalof Weibull, with whom I spoke back in 2002, told me that the seed industry spends more in enforcement costs than it recovers in fines. He claimed that the industry was soon going to correct this situation by setting up a ticketing system. Private investigators would survey farms and issue tickets to violators. The industry would also pursue seed plant operators and cleaners who accepted seeds of protected varieties from farmers.[15] The clear intention here would be to scare local seed-cleaning operators away from handling farmer-saved seed.

In June 2007, I asked Saskatchewan canola farmer Terry Boehm, vice-president of the National Farmers Union, if he knew whether the seed industry was implementing these plans. He said that he has been receiving more and more calls from farmers and

seed cleaners who have been badgered by the CPTA and are confused about their rights. Boehm explained that most canola farmers in the Prairies had abandoned seed-saving years ago, and were
used to buying seed every season. But as of late they had started
to save seed once again because of rising prices. They did not
understand what their rights were under PBR, and the heavy and
somewhat misleading propaganda circulated by the CPTA had left
them even more confused. Plus, the CPTA was stepping up its
enforcement; in the fourteen months preceding January 2004,
CPTA says it won 145 cases against violators of PBR.[16] Boehm
pointed to an ensuing "litigation chill," with farmers and seed
cleaners backing away from saved seed for fear of drawing big law
suits.

Whether it is useful or not, seed companies have never seen
the *Plant Breeders' Rights Act* as their ultimate form of monopoly
protection. For them it has always been just a stepping stone, a
basic legal framework that they could gradually tighten over time,
always moving towards full patent rights. The *Act* does not prevent farmers from saving seeds, or breeders from doing further
breeding with protected varieties. But in 1991 UPOV adopted a
new *Act* that strengthens the rights of breeders considerably. The
UPOV 91 Act had a number of important changes:

- Breeders have rights over the harvest of protected varieties. If
the farmer sows a field to a PBR variety without paying the royalty fee, the breeder can claim ownership of the output (e.g.,
wheat) and the products of the output (e.g., wheat flour).
- Breeding using protected varieties is restricted. Anyone using a
PBR variety in creative research has to make major changes to
the genotype, or else the "new" variety will not be considered
"new" – it will be considered an "essentially derived" variety,
with ownership falling to the first breeder.
- The 1991 Convention does not protect the rights of farmers to
freely use their harvest as further planting material, leaving it
up to individual countries to make special provision for it.

With these changes there is hardly anything separating *UPOV 91* from patents.

Canada is a signatory to *UPOV 91* but is under no international obligation to ratify it. Article 1701 of the 1994 *North American Free Trade Agreement* (NAFTA) specifies that the parties will, at a minimum, enforce *UPOV 1978*, but it goes no further. Yet the seed industry and the Canadian government consistently argue that the absence of *UPOV 91* legislation puts Canada at a competitive disadvantage and at "risk of losing investment and trading opportunities."[17]

In 1998 the federal government introduced a bill to amend the *Plant Breeders' Act* and bring it into conformity with *UPOV 91*. The bill died on the order paper, but the government has since been working with industry to reintroduce the proposed amendments. The original bill, like *UPOV 91*, placed restrictions on further breeding; gave breeders rights over harvests and the exclusive right to "condition propagating material of the plant variety for the purpose of propagating the plant variety"; and limited farmers' rights to "the use of harvested material of the plant variety grown by a farmer on the farmer's holdings for subsequent reproduction by the farmer of the plant variety on those holdings."[18]

Even though the Canadian Seed Trade Association (CSTA), the main lobby group representing seed companies in Canada, approved of the 1998 bill, it is asking the government to go further still. The CSTA says it "is strongly against any farmer's privileges going beyond the provision of the 1991 *Act of the UPOV Convention*, i.e., within reasonable limits in terms of quantity of seed and species concerned and subject to the safe-guarding of the legitimate interest of the breeders in terms of payment of a remuneration and information."[19] In other words, the industry would be willing to let farmers save small quantities of seed, as long as the farmers pay the companies royalties every time they do so.

In the May 2004 report of the Seed Sector Review, conducted

by the Canadian Seed Trade Association and the Canadian Seed Growers Association, and supported by Agriculture and Agri-food Canada, the seed industry renewed its call for Canada to move towards *UPOV 91*. Shortly thereafter the Canadian Food Inspection Agency put forward proposed amendments to the *PBR Act* that would bring it into conformity with *UPOV 91*, and initiated a sixty-day public consultation process on its website. The government intended to bring the amendments quickly before Parliament, but the unexpected public outcry expressed in the consultation, and the context of minority government, put the process on hold, where it remains. One CFIA bureaucrat within the PBR office complained to a member of the National Farmers Union of "literally hundreds" of letters, faxes, and e-mails commenting on the proposed changes having poured into her office.

Step Four: Regulations

The *Canada Seeds Act* of 1923 was initially established, in large part, to protect farmers from seed companies selling poor quality seeds. It instituted a system of mandatory variety registration based on strong performance standards. Although the *Act* has been seriously weakened as an instrument for protecting farmers from the seed industry, and faces a looming overhaul, it still requires plant varieties from most major agricultural crops to be tested for merit (agronomic performance, disease resistance, and end-use quality), and it only permits those varieties that are at least equal to the best available varieties to be sold on the market. The variety registration system is admittedly a holdover from the public seed system, and suffers from its limitations: committees of "experts," composed primarily of formal plant breeders and scientists, commercial seed growers, and commodity group representatives, make the final decisions; the "merit" criteria are biased towards industrial agricultural systems; and there are no mechanisms to specifically assess GM varieties.

In 1998 the Canadian Food Inspection Agency launched a process to overhaul the variety registration system when it commissioned the FAAR Biotechnology Group to conduct a review. FAAR recommended sweeping changes, including the elimination of agronomic merit data in the assessment of varieties. The Canadian Grain Commission explained: "The future role of variety registration for most crops would be solely to recognize varieties from a varietal identity and purity . . . standpoint." A narrow range of health and safety aspects would also be taken into consideration, but the merit criteria would be dropped. "In the absence of prior merit assessment, the actual variety performance (agronomic, quality, disease) would be commodity driven, similar to what exists in the United States," said the Commission.[20]

A CFIA proposal from 2002 aimed to decrease the number of recommending committees recognized by the CFIA from twenty to six.[21] Under the proposal, certain crops – wheat, canola, barley, rye, triticale, oat, mustard, pea, and sunflower – would continue to be tested for merit, but the criteria would include only quality and/or disease resistance, plus a one-year minimum of performance information, down from the three years of tests presently required.[22] This is as good as dropping the merit criteria entirely. With only one year's worth of tests, it is practically impossible to determine the agronomic performance of a variety. According to Rob Graf, a research scientist with Agriculture and Agri-food Canada (AAFC): "The problem with this type of system is that for yield and some other agronomic traits, environment has tremendous influence, which means that one year of data cannot provide a reliable prediction of long-term performance."[23] The CFIA was supposed to institute the changes to the variety registration system in the spring or summer of 2003, but this was shelved for several years, until in 2006, after consultation with seed industry stakeholders, the CFIA came back with a brand new proposal entitled "Proposal to Facilitate the Modernization of the Seed Regulatory Framework," in which it recommended an even greater erosion of the merit criteria. In October 2006 the CFIA launched a sixty-day, Internet-based public consultation process on the proposal to, in its words, "gauge the level of support for the proposed changes."

With a number of crops, testing requirements for agronomic merit were already significantly reduced. The goal was to ease the regulatory burden for seed companies so that they could deliver new varieties to farmers more quickly. But in 1996, SeCan, an association of Canadian seed growers, was already reporting problems: "Often there is not enough data available to make an informed evaluation of a variety, and this problem is getting worse each year as less money is available for testing."[24] Again, in June 2000, in response to concerns from its members about the quality of the varieties it was distributing, SeCan blamed the regulatory system: "We try to release only new varieties that offer benefits over current varieties, but it is often difficult to evaluate varieties from the limited amount of data available. . . . In Ontario, soybean lines can be considered for registration after only one year of testing."[25]

At the end of the 1990s, the number of variety trials required for registering canola was reduced from three to two. Soon after, in 2002, a number of farmers in the Prairies were hurt by the failure of a new canola variety. Variety 45A77, a Clearfield herbicide-resistant canola developed by Pioneer Hi-bred and marketed by Proven Seeds, a subsidiary of Agricore United, was badly damaged by herbicide spray in various fields across the Prairies. Agricore United's CEO Brian Hayward suggested to the *Farmers' Independent Weekly* that the problem may have stemmed from "variable conditions," and he admitted that there was widespread but variable damage. With the dismantling of the registration system, canola does not have to be tested under variable conditions, leaving some to wonder if the system is inadequate. According to John Morriss of the *Farmers' Independent Weekly*:

For the past several years, the mantra for canola registration system is that less regulation is the best regulation and, "Let the marketplace work." Just what that means for the farmer is not so clear.
Canola seed is not like a home appliance with a one-year money
back guarantee. You can't take the seed back if it didn't perform as
advertized. The marketplace may be working to the extent that no
one is likely to buy 45A77 next year, but that doesn't do anything
for farmers looking for compensation this year. . . . The rush to
develop and market the latest "new and improved" variety may be
leading to unacceptable risks for not only farmers, but for the companies who develop and market the seed. There is enough uncer-
tainty in agriculture these days. If steps can be taken to reduce
risks of this magnitude, a little more regulation (or maybe that
should be "protection") may not be such a bad thing.[26]

When the *Plant Breeders Rights Act* was up for consideration
in the early 1970s, public plant breeders voiced concerns that it
might cause the system of merit testing to "break down completely."[27] These concerns were reiterated when the federal government put PBR back on the table in the 1980s. The federal government is on record as having assured them of the contrary.[28]

Merit criteria were once the core of the *Seeds Act* – a guarantee of high-quality seed for farmers. With these criteria now
essentially stripped away, a new criterion has gradually and rather
silently taken their place: that of seed "purity." The first move in
this direction dates back to 1973, when the *Seeds Act* was
amended to discourage farmers from buying farm-saved seed
from each other. It prevented them from referring to their crops
by variety name if they were not grown with certified seed. The
assumption here was that certified seed is superior to farmer-
saved seed – that crops must deteriorate or lose their purity in
subsequent generations.

But why should governments be so concerned with plant
"purity"? After all, what does seed purity mean? For the seed

industry, purity appears to refer to the seed's approximation to certified seed. But this has no direct bearing on plant performance.[29] Moreover, even if such "purity" was a legitimate objective, there would still be little basis for legislating against farmer-saved seed. Most major food crops in Canada are self-pollinating, and farmers can generally save seeds from year to year from these crops without causing any serious diminishment of quality or performance. Canada is renowned for the quality of its wheat, yet even the seed industry acknowledges that over three-quarters of the wheat seed sown in Canada in any given season is farm-saved.[30] Only hybrid plants, which are bred in a particular manner to prevent further breeding and seed-saving, degenerate significantly in subsequent generations. Seed "purity" is therefore simply a technical matter of making sure that seeds are properly selected, cleaned, and stored. Farmers have been quite successful at doing this themselves over the years.[31]

In contrast, the seed industry and government have not taken any steps to protect the "purity" of the Canadian seed system from the introduction of GM strains, despite the apparent problems that such contamination can produce. Consumers in Europe and Japan, two of Canada's most important agricultural export markets, have largely rejected GM foods. This has led to the loss not only of certain markets for Canadian farmers growing GM crops, but also, in certain instances, of export markets for farmers whose non-GM crops were contaminated by GM strains. Saskatchewan farmer Percy Schmeiser even lost fundamental ownership of his crop by way of contamination.

When a GM crop is widely introduced, contamination is inevitable unless precautions are taken. But even with precautions, a certain level of contamination will occur, either by mixing during grain handling, cross-pollination, accidental release, or the persistence of GM crops in fields.[32] This is particularly the case with canola, which has consistently been the biggest GM crop in Canada in terms of acreage. GM canola is ubiquitous on the

Prairies, even in places where it has never been deliberately sown. Saskatchewan farmer Robert Stevenson had never planted GM canola but nonetheless found GM canola plants on his fields. "It's close to being as thick as a crop," he says. "Crop insurance considers nine plants per square metre to be a viable canola crop. Without even trying I have four [GM canola] plants per square metre. This for me is a new weed, and it's here in very significant numbers."[33]

Contamination is happening not only in farmers' fields. Studies show that the certified canola seed supply is deeply affected.[34] Canola breeder Keith Downey suspects that "there are varieties of certified seed out there, in which part of the level of contamination is coming right from the breeders' seed."[35] A study commissioned by AAFC, leaked in June 2002, confirmed the severity of the contamination of canola. The study found that the " . . . large number of canola seeds normally planted per acre plus the high probability that a small percentage of herbicide tolerant seeds will be present in most Certified Seed lots has and will continue to result in significant herbicide tolerant plant populations in most commercial canola fields."[36]

Similar reports are emerging from the U.S. In February 2003 Walter Fehr, an agronomist and director of the Office of Biotechnology at Iowa State University, publicly stated that, in his opinion, there was probably widespread GM contamination of the seed supply for crops where GM varieties have been commercialized – soybeans, cotton, corn, and canola.[37] A year later, a study by the Union of Concerned Scientists confirmed Fehr's predictions, indicating that, in the U.S., non-GM varieties of these crops are "pervasively" contaminated with low levels of GM material from GM varieties.[38]

In this context, co-existence of GM and non-GM crops is not possible. When a GM crop is introduced, it inevitably contaminates other varieties. As it is, the CFIA can only refuse to authorize a GM variety if it can demonstrate that the variety presents certain environmental and health risks. Under the existing regulatory framework for GM crops, which is based on the notion of substantial equivalence, there is no scope for refusing to register varieties modified with today's commercially available GM traits, such as the Roundup Ready or Bt varieties. A GM crop cannot be prohibited on the grounds that it can cause economic losses for farmers, or other socio-economic problems through contamination.

The Canadian government could use Canada's variety registration system for such purposes. When the first GM varieties came through the system, the evaluation committee took the unprecedented step of awarding bonus points for herbicide resistance (the varieties probably would not have been approved otherwise).[39] Now that certain negative implications of GM crops are apparent, conceivably the committees could deduct points from GM varieties where there are negative consequences for farmers. So far, however, the government has not moved on this suggestion, preferring instead to look at other possibilities that cater to the interests of transnational agribusiness.

The government and the seed industry's strategy for dealing with GM contamination, although not formally announced, appears to involve the development of voluntary segregation systems based on "Identity Preservation." Identity Preservation systems are supposed to "preserve the identity of specific lots of grain from farm to market," and give Canada a "significant competitive advantage."[40] The Canadian Prairies, however, already have a system to protect Canada's so-called "competitive advantage." The variety registration system and the Kernel Visual Distinguishability system, whereby grain operators look at batches of grain and decide what class they fall into, are designed to work together to maintain the quality of Canadian exports, and guarantee farmers premium prices on the world market. These systems are the cornerstones of the Canadian Wheat Board, a farmer-controlled organization with single-desk selling authority for wheat and barley grown by Western Canadian farmers. The actual problem for many farmers is not to secure top prices on the world market, but rather to prevent the loss of markets due to competitive disadvantages caused by the introduction of GM and low-quality varieties. The identity preservation systems that the government and agribusiness are proposing would only exacerbate this problem.

These preservation schemes will allow more private sector varieties on the market, both GM varieties that are rejected by export markets, and conventional varieties that do not exceed the standards set by public varieties. Moreover, the spread of identity-preserved production, by breaking with the Kernel Visual Distinguishability system, would undermine the Canadian Wheat Board. It would allow global grain traders like Cargill and ADM to increase their control over the Canadian grain trade, and transnational seed companies such as Monsanto and Syngenta to take over the seed market for crops like wheat, traditionally dominated by public-sector varieties. Halfway through 2007, with a minority Conservative government in power that has promised to terminate the Wheat Board's single-desk selling authority, both the Wheat Board and the Kernel Visual Distinguishability system are on the ropes. If one falls the other will soon follow.

Identity preservation would also shift the costs and responsibility of GM contamination onto farmers growing non-GM crops. As Bill Toews, a wheat farmer from southern Manitoba, points out, "What [the identity preservation system is] trying to do is introduce a lower-value variety [the GM variety] into a stream that has a relatively higher value." This, says Toews, will "add a segregation cost which will be shifted from the GM crop to the non-GM crop, because it is a higher-value crop that we are trying to protect. Why [as farmers] do we want to do that?"[41]

One of the main reasons for the seed industry's interest is that identity preservation systems are an effective way to stop farmers from saving seed, particularly for the cereal crops on the Prairies, where there is still a high level of seed-saving. Farmers growing identity-preserved crops often have to sign a contract that prohibits them from saving and planting seeds from their harvests. In January 2004, *Germination*, the magazine of the Canadian seed industry, asked Canadian "seedsmen" to name the best options for stopping seed-saving. The most common answer was hybridization followed by identity-preserved production, with PBRs far down the list because of difficulties with enforcement. It is not surprising then that the seed industry has pushed wherever it can to expand identity-preserved production and to ensure that these systems continue to be based on contracts that prohibit seed-saving.

The Canadian Seed Trade Association (CSTA) has even developed a model for identity preserved production, which it calls an "Affidavit System." It requires farmers, when they drop their harvests off at grain elevators, to sign a written guarantee attesting to the variety of their crop. In this way, the grain is supposed to be segregated by maintaining the "identity" of the variety through the grain handling system. The merits of such a program are questionable, given the level of contamination in the certified seed supply. If adopted, the Affadavit System would have major implications for seed-saving practices.

The push for such affidavits is tied to an earlier modification to the *Canada Seeds Act*, prohibiting farmers from referring to their crops by variety name if their crops are not grown with certified seed. According to a January 2003 position paper by the CSTA on the Affidavit System:

> A legal opinion obtained by the CSTA confirms the reality that only crops planted with [certified] seed can be identified by a variety name in the grain handling and processing system. . . . We recognize the concerns of industry stakeholders with mandating the use of certified seed. Where products are to be sold by "class," the CSTA supports a middle ground position of not requiring the crop to have been planted with certified seed. However, the grower

must be able to prove the purchase of certified seed of that variety in recent years. In cases where the grain handler or processor is claiming the grain is identity-preserved the requirement for the use of [certified] seed must be complete.[42]

The CSTA's proposal is based on a peculiar interpretation of responsibility. Grain handlers have been sorting farmer-saved seed by class without difficulty ever since the classification system began in Canada. Why should farmers now have to prove that they have used certified seed? Moreover, farmers have generally had little trouble in preserving the genetic "identity" of their crops while saving seeds. Farm-saved seed can cause agronomic problems if the seed is not properly handled, but this will not undermine its quality for the end-user – unless, of course, the crop is at risk of contamination from GM crops. But the seed industry, not the farmer, is responsible for this. It is rather unfair to penalize farmers, by making them buy seed every year, for a problem created by those selling the seed. This is especially true when the certified seed supply is as seriously contaminated as are the farmers' fields, a problem that the seed industry itself admits to.[43] Nevertheless, despite the obvious weaknesses and self-interest in its arguments, the seed industry is gradually getting its way.

In June 2003 the Canadian Grain Commission launched a voluntary program to oversee and officially recognize "identity-preservation" programs in Canada. The Canadian Seed Institute, a "not-for-profit, industry-led organization" founded by the CSTA and the Canadian Seed Growers Association, and managed by a board of industry representatives, is the first agency accredited by the Canadian Grain Commission to offer auditing services for this new program.[44] The Canadian Seed Institute's official involvement in this area dates back to November 2001, when AAFC Minister Lyle Vanclief allocated $1.2 million to the Canadian Seed Institute to develop a "Market Delivery Value Assurance Program," whose goal was to "help develop standards and audit procedures, as well as to launch a research program to verify grain purity, develop an internet-based tracking system requiring key information during each step of the handling process, and create a national third-party certification body." Not surprisingly, the seed industry's proposals are integrated into the Standards of the Canadian Grain Commission's identity preservation program. Section 5.4.2 states: "The company shall ensure that appropriate stock seed is selected to fulfil the IP contract, and that the seed is traced to the grower. Where the IP contract is variety specific, certified seed shall be used."[45]

Such variety-based contracts will become more widespread if the CFIA's "Proposal to Facilitate the Modernization of the Seed Regulatory Framework" goes through. Under a special clause in the existing seed regulations dating back to 1996, certain high-value crops that could not pass the requirements of the variety registration system, and that cannot be comingled with ordinary crops and commodities, can receive authorized contract registrations setting out strict segregation requirements. As of 2006, only nine varieties have been grown in Canada under contract registrations. The CFIA's proposed changes, developed through consultations with the seed industry, would shift the criteria to a "risk-based assessment," making it much easier for seed companies to qualify for, and adhere to, contract registrations. Plus, varieties under contract registrations will not have to pass through the merit tests required of other varieties.[46] The seed industry, therefore, would have every reason to introduce more and more of its new varieties through contract registrations, leaving farmers fewer and fewer options for buying seeds that can be saved.

Identity preservation systems are really just a form of contract production in which the farmer has almost no autonomy. They are no longer a marginal element of Canadian agriculture, and, as the global agri-food industry continues to restructure towards vertically integrated supply chains, they could well become the norm. In January 2004 *Germination* magazine also reported that nearly

two thirds of the Canadian seed company representatives it surveyed believed that over 30 percent of the market for seed for all
crops will be under identity-preserved production by 2010. More
than a quarter of them stated that over half of the market would
be identity-preserved by that time. The changes to the regulatory
system that have already occurred, and those that are being considered, are indeed designed to facilitate such a development.

In 1992 William Leask, then executive vice-president of the
Canadian Seed Trade Association, put forward a three-wave theory
of crop development. His second wave corresponds to the post-
World War II public seed systems, and his third wave is based on
identity preservation systems. According to Leask: "Merit has traditionally been defined by the needs of farmers, such as yield
improvements and disease resistance. . . . The third wave means
that merit will be determined by the crop's utility further down the
chain of production."[47] Of course, this "utility" will be tightly controlled, and owned, by the seed industry and the larger corporate
structure or supply chain it belongs to, and the state will be called
upon to ensure this control and ownership. According to Jeff
Guest, chair of the Variety Name Preservation Working Group of
the Canadian Seed Trade Association :

> We have these rules [the *Seeds Act* and Plant Breeders' Rights]
> that protect us on the seed side, but nothing as clearly defined for
> the downstream processing side. . . . As we go forward in the next
> decade, more people in the agri-food chain are going to be
> involved in identity preservation. As that's happening, we need
> intellectual property regulations to keep pace with where industry
> is going to ensure protection for seed companies, along with intel-
> lectual property protection for everyone in the identity-preserved
> chain.[48]

Things are moving in this direction. There is an ever-growing
number of tie-ups between seed corporations and downstream

grain merchants and food processors; for example, the Renessen joint venture between Monsanto and Cargill, or the Solae joint venture between DuPont and Bunge, one of the world's largest soybean traders and processors. In December 2006 Cargill and Bayer announced that they were teaming up to launch a GM canola variety in 2008 in Canada, with a "specialty oil" trait and resistance to the herbicide glufosinate. Cargill will exclusively handle the marketing of the seeds and the crops through an identity preservation program. A Saskatchewan-based seed and agribusiness company, FarmPure Foods, which was selling only seeds a few years ago, now markets ready-to-eat oat products and a gluten-free beer called NuBru, made only with its PBR-protected plant varieties. "NuBru is what our vision for building value chains is all about," says CEO Trenton Baisley.[49]

✺ ✺ ✺

In the decade or so that I have spent following transnational seed corporations, I have never ceased to be amazed at the lengths to which they will go to advance their interests. To give just one example from outside of Canada, my colleagues at GRAIN and I recently looked into a spate of new seed laws sweeping across Africa. These laws were essentially written up by seed corporations and their friends in UPOV and other international institutions. On a continent where at least 90 percent of the local food is produced with farm-saved seeds, it is shocking that the vast majority of these new laws are targeted at preventing the use of farm-saved seeds and the traditional seed systems that sustain them – all in the name of building up big enough markets for certified seeds to justify corporate investment. The laws would be genocidal if not for the incapacity of these governments to implement them fully.

Few African farmers have any inkling of what the seed companies and lawmakers have concocted for them. They continue on as before, unaware of the precariousness of practices they have always taken for granted. The same goes for farmers in most industrialized countries as well. People from the private seed sector in Canada regularly express their dismay at the stubborn refusal of farmers to stop saving seeds. But maybe it is just that most farmers have a tough time getting their heads around the abstract idea that the seeds, which come from the plants they have carefully cultivated and grown on their own lands, no longer belong to them.

Seed corporations are constantly ramping up the pressure, trying to get the message across to farmers with ever-more creative and agressive schemes. "Canada is simply not a profitable place for private breeding institutions to invest," says FarmPure Foods CEO Trenton Baisley. "Brown-bagging is so entrenched; we need new mechanisms that encourage farmers to invest in innovation, market opportunities, and the use of certified seed." So Baisley came up with the idea, now championed by the CSTA and the Canadian Association of Agri-retailers, of a certified seed subsidy. In the carefully chosen words of John Cowan, past-president of the CSTA, it is "a tax credit that would help farmers offset the increased cost of certified seed."[50]

Then there is the idea of tying crop insurance to the use of certified seed. Seed companies have lobbied for this for years, although, to date, the only province where farmers cannot get crop insurance without purchasing certified seed is Quebec. Terry Boehm of the National Farmers Union suspects that this could change soon on the Prairies. When I spoke with him in June 2007, he said that insurance companies have started demanding detailed accounts of the varieties farmers are planting, before providing them with crop insurance. He worries that it may not be long before they start insisting on certified seed. In Quebec, however, disgruntled farmers are taking action to get around the rules. In 2005 I was invited to a meeting in the town of Sainte-Hyacinthe, not far from Montreal, organized by organic farmers developing a protocol for farm-saved seeds that they could use as a way to certify their own seeds as organic, and qualify for crop insurance. The farmers there spoke about how certified seed was generally ill-suited to their farms and their organic production methods, and how, by saving and selecting seeds from their own fields, they could have much better success in developing crops adapted to local conditions. The idea of the protocol is still alive among these farmers, even though, given the minimal support from governments, the antagonism of the provincial seed growers, and the day-to-day pressures of keeping a small organic operation alive, they are struggling with the politics and practicalities of putting it in place.[51]

Rumblings of resistance to the seed industry's shenanigans can be heard in other parts of Canada as well. Early in 2007, farmers with the Manitoba Canola Growers Association, upset with seed companies not supplying the varieties they wanted on reasonable terms, put forward a resolution at their annual general meeting calling on the association to lobby federal agencies to allow farmers to save seed of varieties not adequately supplied by seed companies, even if these varieties were tied up in patent rights. The move flows out of frustrations that farmers in the Prairie provinces are having with a new trick that pesticide/seed corporations are using to stop them from saving their seed. These companies have ceased making their leading canola seed treatments – like Syngenta's Helix and Bayer's Prosper – available, except for use on their patented seeds. Bayer's Gaucho seed treatment was, for a time, the only one still available to independent seed cleaners, but in 2005 the company started requiring seed-cleaning plant operators to sign user agreements obliging them to prove that they were not applying Gaucho to "illegal brown-bagged canola seed." Many small, co-operative seed-cleaning plants responded by refusing to treat farmers' canola because they had no way of knowing for sure whether the seed brought in for treatment was patented or PBR-protected. With those seed-cleaning plants that signed on to the agreement, Bayer followed up with audits to see which farmers were bringing in their own seeds. Other plants, which refused to sign the agreement, simply stopped treating canola because they didn't have the necessary chemicals. In 2006 Bayer went a step further and started requiring farmers bringing in their seeds for treatment to sign user agreements as well.[52]

"We created a system for cleaning seed for farmers at an economical cost for both farmer and us and they're going to start forcing our seed cleaning plants out of business and I think that's terribly, terribly wrong," said Mel Foat, manager of the Sexsmith Co-operative Seed Cleaning plant in northern Alberta. "When chemical companies are starting to get in the seed business and starting to rule and dictate, if Bayer thinks this is going to stop the growth of F2 generation seed, it's just a joke." Foat believes these moves are going to force farmers to mix their own illegal cocktail of seed treatment, and to revive old fanning mills to clean their seed.[53]

Others, like Terry Boehm of the National Farmers Union, think that by going after seed cleaners the seed industry could stop a lot of farmers from saving seeds. "It's geared towards what's absolutely critical to farming out here in Saskatchewan," he says. "Farmers won't plant seed without cleaning it."[54]

Bill Ross, executive manager of the Manitoba Canola Growers Association, points out that there is nothing illegal about saving non-patented, open-pollinated canola seed varieties such as the Polish and Argentinian varieties that farmers still commonly grow in the province. "The problem," he says, "is that they have nothing to treat it with."

With rising prices for seeds and not much in the way of innovation, Canadian farmers, particularly those on the Prairies, are struggling to find some way to loosen the grip of the seed companies. Seed growers in Alberta are proposing a check-off fund, based on levies from seed growers and contributions from the provincial government, to support plant breeding that would develop farmer-owned varieties. These could be made available royalty-free. Seed growers in Manitoba are also mulling over the idea. "The only way that the seed growers endorsed it – supported plant breeders' rights – was if the government maintained or increased their research dollars spent in agriculture," said Dauphin, Manitoba seed grower Rod Fisher. "The day plant breeders' rights came in, the (government) research dollars started to be cut . . . and now we're back at the government to give us some of the money they were supposed to give us all along."[55]

Meanwhile in Wawanesa, Manitoba, a group of canola farmers decided to take seed research into their own hands. In 2004 they started testing the performance of farm-saved, hybrid canola seed. Hybrid seed is supposed to lose its "hybrid vigour" in subsequent generations, and therefore, farmers are not supposed to be able to save seeds. This is why, with canola, private seed companies pretty much offer only hybrid varieties. But the Wawanesa farmers were not convinced, and they decided to conduct a trial on a sixty-acre field. The outside rounds of the field were planted to certified hybrid seed, while the rest were sown to the common seed in order to get a side-by-side comparison. The farmers say that there was virtually no difference in yield, and they suspect that the hybrid seeds weren't really all that different from conventional seeds to begin with.

"If the parents are distinct, then you could run into problems. I think the reality is they're not always distinct," said Warren Ellis, one of the Wawanesa farmers. "The thing is that you'll never find out whether they're distinct or not. I don't know how you'd ever find out. I don't suppose the companies are going to tell you."

Ellis says that farm-saved seed costs about five times less than certified, hybrid canola seed. "We're not scientists here," Ellis said. "We're just lay people trying to figure how to make money farming."[56]

The fight for farmed-saved seed is far from over in Canada. The worst from the seed corporations is probably yet to come, if not in the form of Terminator seeds, then through other equally or more disruptive mechanisms. In Brazil, one of the great experimental grounds for seed policing, companies are already setting up a system to monitor farms by satellite to ensure that only certified seed is being grown.[57] But to this day, the seed industry, with all its political connections and heavy economic clout, has still not killed the spirit – or might it be the necessity – that keeps many farmers saving seeds. When the CSTA asked seed company representatives what needed to be done to stop farmers from brown-bagging in Canada, one of them candidly answered that you would have to shoot them.[58]

Chapter 7

Privatization

It is no secret to anyone in the seed business that private seed companies cannot make big profits when they have to compete with public varieties. Public varieties do not cost farmers as much, because the costs of research are generally paid through public funding rather than the price of the seed, and because farmers are free to save the seeds from year to year. J.A. Stewart of Alex M. Stewart and Sons, a small Ontario seed company, laid it out bluntly for the Canadian public and private plant breeders assembled at a conference in 1971: "[There must be] fewer public sector breeders and fewer public varieties, if seed companies are to survive."[1] But back then, hardly anyone involved in plant breeding seriously entertained the idea of dismantling the public sector breeding programs.

So when, in the 1970s, some people within industry and government started to advocate more openly for private sector investment, their efforts were always scrutinized through the lens of the public programs. What impact would such moves to encourage private sector investment have on the public breeding programs? And among public plant breeders, and even within government, there was a frank and open recognition of the inherent contradictions between a public and a private seed system. C.R. Phillips, the director general of the Production and Marketing Branch of Agriculture Canada, told the same 1971 conference: "The most significant incentive for private breeding is the cessation of public breeding or for the public breeder to act like a private breeder and charge sufficiently for the seed to recover cost." Governments, therefore, had to make a choice between supporting public or private plant breeding; you couldn't have it both ways. Here, Philips was clear about what choice the Canadian government should make: "I believe it would be very difficult to demonstrate that private breeding would be superior to public breeding, particularly when you consider . . . the particular climate, crop, and acreage conditions in Canada," he said.[2]

In 1984, Consumer and Corporate Affairs Canada commissioned R.M.A Loyns and A.J. Begleiter of the University of Manitoba to produce a working paper on the potential economic effects of PBR, which the federal government was once again considering to implement. Loyns and Begleiter studied the experiences of other countries with plant breeders' rights and surveyed a large number of public plant breeders and representatives of seed companies from Canada and other countries. They concluded that PBR were unlikely to have much of a positive, or negative, impact on plant breeding in Canada. PBR, they argued, do not have any significant advantages over the current system. Among those surveyed,

> [t]he feeling exists that the present varieties licensing system . . . could fulfil all the domestic requirements of PBR. That is, royalties can be collected . . . on new varieties if the breeder or breeding organization so chooses. The licensing system ensures that the new varieties are visually distinguishable from existing ones as well as meeting all existing quality standards and exceeding at least one of them. In this latter regard, the Canadian licensing system imposes a more stringent requirement than the UPOV system. . . . [It was also felt that] the current system of individual agreements between foreign seed companies and their Canadian counterparts would allow Canadian farmers access to the best foreign varieties.

Moreover, "there was unanimous agreement among plant breeders that there had been no change in the rate of [seed] exchange with breeders in countries which had adopted PBR."

According to Loyns and Begleiter, PBR, in the view of both the public breeders and seed industry representatives they contacted, were not going to make a significant difference to the Canadian seed supply.

Yet, the report did uncover a great deal of anxiety about PBR among Canadian plant breeders:

> Despite federal assurances to the contrary, a good deal of concern was expressed in both the public and private sector about the possibility of reduced government support for public plant breeding in the future, especially for varietal development. . . . The majority concern in Canada seems to be that the federal government is introducing plant breeders rights in the expectation that increased private sector investment will allow it to decrease its support for plant breeding. There is support for the idea that SeCan already provides most of the protection plant breeders' rights is intended to provide. . . . Also, although the federal government is on record as being committed to maintaining current variety licensing requirements, quite a number of people expressed the view that if private investment in plant breeding increased, there would be irresistible pressure brought to bear to modify the licensing system.[3]

Echoing the sentiments of many public plant breeders, Keith Downey, AAFC's most well-known plant breeder, told his Canadian colleagues in 1971: "Exchange of vital and important genetic material at a very early stage is also part of today's scene and is based on the belief that such exchanges will be reciprocated. . . . The walling off of certain areas or crops for public breeding while leaving the rest exclusively for private breeders will not work." Downey was concerned that intellectual property rights, particularly patents, would encourage the breakdown of merit testing for new varieties, the loss of the public sector's team approach to plant breeding, and the promotion of foreign varieties "of questionable adaptation and performance in the face of equal or superior Canadian public varieties."[4]

Despite such concerns, a change in perspective began to take root within the federal government. The honest acknowledgement of the brutal impacts that PBR and any other mechanisms to support seed companies would have on public programs gradually gave way to an idealistic effort to support a private seed system that would "complement" the public sector.[5]

Enter Monopoly Rights

The first change of significance occurred in the mid-1970s, with the creation of the SeCan Association. SeCan is an association of seed growers, much like the Canadian Seed Growers Association, except that it charges a membership fee and allows seed distributors, processors, and others involved in the seed market to join. There was another, more fundamental, difference between the two seed associations. SeCan made agreements with public breeding programs for *exclusive* licences to multiply, distribute, and market varieties. Only SeCan members could grow varieties licensed by SeCan. SeCan charged a levy of 2 percent on the sale price of certified seed, and collected any royalty that the plant breeder chose to impose. Agriculture Canada, which at least until the early-1980s undertook roughly 70 percent of the breeding work, chose not to collect royalties from the varieties it licensed to SeCan, perhaps recognizing the inherent contradiction in the idea of a public institution generating royalties from public varieties. Nonetheless, a precedent was set. By charging levies and collecting royalties for breeders, SeCan shifted some of the costs of plant breeding from the government to farmers, signifying a new perspective on plant breeding as a business with farmers as the customers, rather than plant breeding as a national activity carried out in collaboration between breeders and farmers, with Canadians and Canadian industry as the beneficiaries.

In the 1980s the federal government began to cut back the budget for public plant breeding programs and introduced changes to how funding was allocated. Although precise figures for federal expenditures on plant breeding are not available, economist Phil Pardey claims that public research expenditures for agriculture declined from 8.4 percent of overall public spending on research and development in 1981, to 2.9 percent in 2000.[6] Moreover, the shrinking proportion of money going into budgets for public agricultural research was accompanied by new funding programs encouraging public researchers to pursue partnerships with the private sector. A large chunk of AAFC's research budget was channelled to the Matching Investment Initiative. Through this program, launched in the mid-1990s, AAFC matches industry investments in collaborative agriculture research projects with public programs. In the fiscal year 1997–8, AAFC spent $64.4 million on Matching Investment Initiative crop research projects.[7] Keith Downey says that this overall reorientation of funding was already affecting his canola breeding program in the early 1990s:

> It used to be that we could say to the outside funders, give us enough to get the hands to run this stuff. We won't worry about supplies or travel, we have that in our basic budget; we just need hands. But then it got to the point where we didn't have enough money in our budget to buy supplies, and keep the place operating, so we had to build that in. Now basically the outside money is running the whole show.[8]

Intellectual property rights are also transforming public plant breeding. The success of Canada's programs is due to a culture of co-operation and the free exchange of germplasm and information among breeders and institutions in Canada and abroad. The way one AAFC plant breeder sees it: "Plant breeding is incremental. We all stand on the shoulders of everyone who has gone before us and add our little bit."[9] A University of Saskatchewan plant breeder makes a similar point: "If you don't give, you don't get, and if you don't get, you're dead. . . . All of the germplasm we use we get from someone else. It takes a whole career for this exchange of germplasm to balance out."[10] Such give-and-take is threatened by the competitive culture of IPRs and royalties.

Followed by Competition and Secrecy

In 1999 Steven Price, a plant breeder with the University of Wisconsin, sent out a survey to 187 public breeders in the U.S., asking them about difficulties they may be having in obtaining genetic stocks from private companies. Forty-eight percent of those who responded said that they had had difficulties obtaining genetic stock from companies; 45 percent said it interfered with their research; and 28 percent said that it interfered with their ability to release new varieties.[11] Here in Canada a germplasm developer with AAFC told me how he has spent nearly ten years now trying to get access to a wheat variety from a private company. He said that, in discussing the matter with his Canadian colleagues, he has come to understand that he's not alone in thinking that "germplasm exchange is slowing down."[12]

Obstacles to germplasm exchange, however, are not coming solely from the private sector. Some public breeders in Canada are already refusing to share their research with colleagues. A University of Saskatchewan plant breeder says that certain breeders at AAFC were working with genes that they had identified for disease resistance. Most breeders would have exchanged their most advanced material incorporating these genes with other breeders, but in this case the AAFC breeders refused even to share the resistance genes. After pressure from other breeders, they agreed to share, but only in the form of raw, early germplasm, making it very difficult for other breeders to work with the material.[13]

It could be said that this has nothing to do with Canadian intellectual property regimes since PBR provide research exemptions. Under the *PBR Act*, the owner of a protected variety does not have the right to restrict other breeders from using that variety. But PBR, like patents, provide exclusive rights and the potential for royalties, and, therefore, they create the incentive for breeders to keep their research to themselves until they have received PBR or patents. Even if public breeders are not interested in going down this road, the people above them are insisting. University plant breeders openly admit that their administrators are demanding that they secure intellectual property rights over their research before exchanging information and genetic material with other breeders.[14]

In 2003 it surfaced that senior bureaucrats in the AAFC were contemplating a policy change that would see their plant breeders personally collecting a portion of the royalties from varieties they developed. Jim Bole, AAFC's science director of cultivar development and genetic enhancement, commented: "I don't know that there is any deadline or if anybody has been specifically asked to come up with this but I do know that it has been discussed from time to time and in fact discussed recently."[15] Whatever the case, the AAFC is already moving aggressively down the patent route in the U.S., where patents on plant varieties are permitted. AAFC has a U.S. patent on a new canola variety that it developed in collaboration with Saskatchewan Wheat Pool.[16] AAFC might argue that the patent is a defensive move to prevent others from patenting its work, but it could just as easily have published its research to keep it in the public domain. A plant breeder with Saskatchewan Wheat Pool says AAFC insisted on applying for a patent: "They were more interested in the potential profit than we were."[17]

This new culture of competition and secrecy in plant breeding obstructs research in more indirect ways as well. A plant breeder from the University of Saskatchewan said it used to be that breeders operated according to an unwritten code of ethics, whereby if you received material from another breeder and discovered something of value within it, you simply got that breeder's permission to carry out further work with the material. There was no legal fussing and people never refused to give permission.[18] Times have changed.

When barriers are put up to co-operation and exchange, innovation in plant breeding is constrained. Proponents of patents and plant breeders rights argue that such constraints are offset by increased private-sector investment, which is more efficient and responsive to market demand. But increased innovation in the private sector is not a substitute for innovation in the public sector. Public breeding programs have different objectives than private ones, and as breeding shifts from the public to the private sector, the outcomes of plant breeding change accordingly, as with canola breeding in Canada.

In 1970, before canola became a major crop, 83 percent of the total research spending on canola was public investment. By 2000 the numbers were reversed, with the private sector accounting for over 85 percent of the total $160 million expenditure. Similarly, before 1973 all varieties were public; between 1990 and 1998, 86 percent of the varieties introduced were from private breeders.[19] Some might point to canola as a success story: solid, public breeding spurs a wave of private investment that allows the public sector to back away. But even if research is only considered from the narrow measure of total investment dollars, the model for canola may not work with other crops. First, canola is unlike many of Canada's other major crops: there is a large market for its seed, and farmers have less incentive to save their own canola seed, although this is changing because of rising seed costs. Second, private investment was encouraged by a large amount of public support and subsidies. And third, canola is attractive to the transnational seed industry because it is an easy crop to engineer genetically. But even if we disregard these unique attributes of the canola seed market, the question remains: Is the goal of state policy the development of a private seed sector, or the improvement of canola? The transfer of research to the private sector has not helped to "improve" canola for the benefit of farmers and other Canadians.

A Change in Priorities

In their study of plant breeding policy in Western Canada, agriculture economists Richard Gray, Stavroula Malla, and Shon Ferguson show that large private investments in canola R & D in the 1980s and 1990s did not significantly augment the rate of increase in crop yields. This private investment did, however, transform the objectives of plant breeding. Private sector investment went primarily into the development of hybrid varieties and varieties resistant to herbicides. By 1999 one half of the canola area was seeded to herbicide-tolerant varieties that required farmers to sign technology use agreements or use specific herbicides. In 2000 over two thirds of canola acreage was either planted under production contracts, or sown with canola varieties dependent on specific herbicides. Moreover, as we have seen, the widespread planting of GM canola has contaminated conventional fields, leading to a situation where it is practically impossible to grow GM-free canola in Western Canada. The first hybrid canola variety was introduced in 1989; by 1997, hybrid varieties had a 30 percent market share.[20]

Gray, Malla, and Ferguson also warn that private investment may actually reduce overall investment in the long run:

> While the current canola research industry appears to be very competitive, there are some concerns for the future level of competition in the industry that are related to the issue of "freedom to operate." Many of biotechnology processes and the genes used in the breeding of canola are patented and have become the property

of many firms in the industry. A single new variety may require three dozen different licensing agreements for the use of use of genetic material and the processes used in its production. The negotiation and the construction and enforcement of contracts to manage this property is a very costly activity. These costs have raised the issue of the freedom to operate for firms in the industry. Because larger firms can more easily deal with these costs, and the costs can be avoided if firms merge so that these transactions take place within a firm, the property rights will tend to accelerate firm concentration in the industry. This raises the spectre of insufficient long-run competition in the industry potentially reducing the long-run investment research and having research products that are sold at higher than competitive prices.[21]

Canola provides an example of what happens when private companies begin to dominate plant breeding for a specific crop. But for other crops with less attractive seed markets, if the public sector backs away, no companies will likely be there to pick up the slack. As one bean breeder from the Ontario Agricultural College notes, in the case of crops like pulses, transnational seed companies are unlikely to carryout region-specific plant breeding. They are much more apt to look within their existing portfolios to see what "cast-offs" from their collections might work in Canada.[22] In reality, the seed industry is only going to invest in plant breeding for crops where there is a potential for large profits. And these profits are only going to flow to the industry when the public sector is shut out of applied breeding, when there are higher seed prices and market shares for the big seed companies, and when farmers are prohibited from saving seed.

The Case of Soybean Breeding

There has not been much outright privatization of public agriculture research programs in Canada involving the transfer of programs to corporate ownership. It is more common for public research to be reorganized in such a way that the program continues to be publicly financed, while control over the research agenda shifts to corporations. Public breeding programs have generally been reoriented to areas of research that support, or at least do not interfere with, corporate interests, although such areas are constantly shrinking as the corporate presence in the seed market increases. Ultimately, all of the direct links between public researchers and farmers are broken; the private sector steps in between, using patents, contracts, and licences to control what reaches the farmer, how it reaches the farmer, and at what cost. The transition was particularly sharp with soybean breeding, today a major target for corporate investment, but historically the domain of public breeding.

Corporations have heavily directed their investments in the seed industry towards soybeans. In 2005, GM soybeans were planted on more acres than any other crop – nearly 120 million acres worldwide. But, in Canada, soybean plant breeding has traditionally been a public activity. The shift to the private sector and the corresponding rise in corporate investment in the Canadian soybean seed market has thus dramatically altered the context of soybean breeding in Canada.

The Rise of Industrial Soybean Agriculture

Soybean agriculture in North America began in 1765 when Henry Younge planted several seeds from China on his farm in Thunderbolt, Georgia. Other soybean varieties from China and Japan were introduced over the next hundred years, and by the early part of the twentieth century soybeans had become a relatively popular forage crop in the U.S. In 1924 total soybean production stood at 1.8 million acres harvested.

U.S. soybean production escalated during World War II, when increase in demand for fats and oils occurred as the U.S.

cotton crop was devastated by the boll weevil. Government and industry looked to an alternative edible oil crop. Production continued to increase in the postwar period with the collapse of the Chinese soybean export market in 1949, and most importantly, the rise of specialized intensive livestock operations dependent on the capital-intensive manufacture of soy-maize animal feeds. U.S. soybean production grew from 18.9 million acres in 1954 to over 60 million acres by the 1990s.[23] Soybean and corn, which are often grown in rotation, now alternate as the U.S.'s largest agricultural crops, with the average annual soybean harvest worth well over U.S. $10 billion.

The increased production of soybeans was not confined to the U.S. Worldwide, the area planted to soybeans expanded from 15 million hectares in 1950 to 72 million hectares in 1999, with a world soybean harvest of 159 million tons, nearly ten times the 1950 harvest. Less than one tenth of this global soybean crop is used for food; the bulk of the harvest is crushed to produce soybean oil and soybean meal. The meal was once the secondary product, but because of the strong demand for animal protein, and hence for protein feed supplements, the meal that is left after the bean is crushed to get the oil is now worth more than the oil itself.[24]

The link between soybean and corn production has made soybeans a natural target for the seed industry, and as soybeans are primarily grown for animal feed, they are also logical candidates for genetic modification. The first GM soybeans were introduced in the U.S. in 1996, and by 2002 they accounted for 81 percent of the entire U.S. crop.[25] Since Monsanto's Roundup Ready varieties have not resulted in significant yield increases, the widespread and rapid adoption of these GM soybeans is often attributed to the ways in which they simplify chemical weed control. U.S. agronomists Michael Mascarenhas and Lawrence Busch put forward other reasons. They point out that "Monsanto's 'no-replant' licensing clause, which prohibited seed-saving, combined with the fact that farmers could only use Roundup herbicide on the seed they purchased, meant that the commercial retailers could substantially increase their revenues." Busch estimates that seed-saving for soybean in the U.S. dropped from 31 percent in 1991 to 10 percent in 2001, largely as a result of the "no-replant" clause in GM crops, and he estimates that this translated into nearly U.S. $400 million in additional gross profits for seed companies and their biotechnology partners.[26] It is not surprising, then, that seed companies are focusing on the release of soybean varieties with the Roundup Ready trait. Nor is it surprising that GM soybeans have encouraged consolidation in the seed industry. By 2002, just four companies controlled 49 percent of the global soybean seed market and 53 percent of all patents related to agricultural biotechnology for soybeans.[27]

Soybean Production in Canada

In the 1940s the Canadian oilseed crushing industry began encouraging farmers to grow soybeans to meet wartime edible oil shortages. But because of the plant's particular daylight needs, soybean production was limited to southern Ontario. Public-sector breeding programs in Ontario released several Chinese-derived varieties from the 1930s to the 1950s, but none of these were suitable for the short, northern growing season.[28]

In the 1970s when U.S. embargoes on soybean exports created another round of supply shortages for Canadian crushing plants, the federal government renewed its interest in the development of short-season soybeans. Lorne Donovan and his colleague Harvey Voldeng, both of the Ottawa Central Experimental Farm, were given the task of developing early-season soybean varieties for the Ottawa River Valley of Ontario and Quebec, and for southern Manitoba. They began working with varieties obtained from a Swedish colleague, Sven Holmberg, who was crossing short-season varieties from the Sakhalin Islands of northern Japan with

German varieties. Donovan and Voldeng crossed Holmberg's varieties with some from the U.S. Midwest, leading eventually to the release of Maple Arrow and other successful short-season varieties.[29]

By the 1980s there was significant production of soybeans in Quebec, Manitoba, and the Maritimes. In Ontario, the area planted to soybeans increased from 4,000 ha in 1972 to 61,500 ha in 1982. By 1987 Canada was the seventh-largest producer of soybeans in the world.[30]

The increase in soybean production in Canada also stimulated private-sector interest in the soybean seed market. Private companies began breeding soybeans in Canada in the 1980s, and by the start of the twenty-first century private varieties dominated the market. Some of the Canadian soybean seed companies have been purchased by larger foreign firms, such as Monsanto, which acquired First Line Seeds, and much of the soybean seed now sold is GM, particularly Roundup Ready varieties. Approximately 50 percent of the soybeans grown in Ontario and 28 percent of the soybeans grown in Quebec in 2003 were genetically modified.[31]

The Changing Nature of Soybean Breeding: Perspectives from Some Public Soybean Breeders

Between 2002 and 2003 I interviewed a number of public soybean breeders to get their perspectives on the changes that have occurred in soybean breeding over the past couple of decades. One of the breeders I met with, an AAFC plant breeder who led the public effort to develop short-season soybeans in Canada, began the interview by saying: "Of the crops I've worked with or watched, soybean has been the most open – best exchange of germplasm, of ideas, and of information among the breeders. Corn breeders tend to be paranoid with secrecy. They hide the pedigree of their hybrids. . . . Soybean breeders have been just the opposite."[32] An open culture of exchange of varieties among soybean breeders allowed this plant breeder and his colleagues to develop the first short-season varieties for Canada, and this open culture has remained critical to Canada's soybean breeding programs.

Exchange with U.S. programs is particularly important for Canadian breeders. According to another of the AAFC breeders I met with: "The soybean breeding community, both public and private, is pretty much a North American community."[33] He told me that Canadian programs for early-maturing varieties are always behind the times of programs in the U.S. because disease problems tend to move northward, from zones with later-maturing varieties to zones with early-maturing varieties. While Canadian soybean crops have been relatively immune to many of the serious diseases in the U.S., he felt that "our holiday is over now." Among the major diseases now threatening Canadian soybean production, he singles out the soybean cyst nematode (SCN), for which pesticides are not economically effective. A plant breeder with the University of Guelph agreed. SCN, he explained to me, is creeping up from the northeast of the United States, and every year it is detected in new counties in Ontario.[34]

As diseases tend to move northward, Canadian research into soybean diseases tends to follow U.S. research. This did not matter when breeding in the U.S. was conducted by the public sector, but today the situation is rapidly changing. A breeder with the U.S. Department of Agriculture, and curator of a national germplasm collection, told me that the private sector started getting involved in soybean breeding in the 1970s.[35] He said that, until a decade ago, public and private programs operated according to a co-operative model but, with the takeover of seed companies by larger firms and the move towards patents, things are "very different now." Private companies are no longer working with other programs or making their varieties available for other breeders to use. The same trend is at work in the public sector, where it is "not driven by the scientific community but by legal and administrative arms of universities."

The U.S. plant breeder said he believes that the patenting of plant varieties "makes no sense." He pointed to a recent Monsanto patent on a quantitative trait loci (QTL) in a wild soybean variety. A QTL is a set of DNA sequences, or genes, linked to a specific trait. He says that his program has identified a similar QTL in their collection, and he wonders what this might mean for their ability to use it. He admitted that, at this point, his team has no idea what happens when a company like Monsanto patents material from their collection. As he explained it, "It doesn't make sense to me that you can patent a gene when all you've done is to just find it."[36]

According to him, most of the work on disease resistance still falls on the shoulders of the public breeding programs because the private sector is concerned with other priorities. "That's the division of labour that [private firms] have come to expect." But this has not stopped private companies from patenting some of the most important work done on disease resistance. With a big disease such as SCN, U.S. firms are racing to patent QTLs for resistance. Pioneer Hi-Bred, for instance, has a patent on "positional cloning of soybean cyst nematode resistance genes" (U.S.6162967), which claims isolated and recombinant DNA of QTLs for SCN resistance. Beyond such patents on specific QTLs, firms are also patenting the methods for identifying SCN and other disease resistance genes or QTLs, and even the methods for breeding for disease resistance, particularly with regard to the use of molecular markers (see Table 4).

One plant breeder with AAFC is working to breed resistance for SCN into early-maturing cultivars using southern Ontario and Midwest lines known for resistance.[37] Initially he was using one molecular marker, published and publicly available, but he says this was not very effective because there is more than one gene involved in SCN resistance. He told me that he would like to use

Table 4
Some Patents Related to Plant Breeding
for Soybean Disease Resistance

Firm	Patent
Pioneer Hi-Bred	QTL mapping in Plant Breeding Populations (U.S.6399855)
Pioneer Hi-Bred	Soybean Cyst Nematode Resistant Soybeans and Methods of Breeding and Identifying Resistant Plants (WO9520669)
Monsanto (Dekalb)	Process predicting the value of a phenotypic trait in a plant breeding program (U.S. 6455758)
Pioneer Hi-Bred	Positional cloning of soybean cyst nematode resistance genes (U.S.6162967)
Purdue Research Foundation	Methods for conferring broad-based soybean cyst nematode resistance to a soybean line (U.S.6096944)
Southern Illinois University	Soybean sudden death syndrome resistant soybeans, soybean cyst nematode resistant soybeans, and methods of breeding and identifying resistant plants (U.S.6300541)
DuPont	Method for identifying genetic marker loci associated with trait loci (U.S.6219964)

Source: United States Patent and Trademark Office, January 2004.

more markers but he is worried that, if he does, the plant varieties he develops might be subject to patent restrictions. His inclination would be to avoid intellectual property rights problems by making a further cross, "taking my resistant line and crossing it to another SCN-resistant variety without using marker-assisted selection." The problem, though, as he acknowledged, is that some of the patents in the U.S. now cover progeny, and while "right now you can't patent multicellular life forms" in Canada, he is worried about infringement if the varieties go to the U.S. As a result, he tries to avoid using markers and methods patented in the U.S.

These concerns are shared by the plant breeder with the University of Guelph.[38] He was working with molecular marker techniques to search for and incorporate SCN-resistant genes from three wild soybean (*Glycine soja*) plants that he obtained from the U.S. germplasm collection. In his breeding, his group was using simple sequence repeats (SSRs). "The beauty of SSRs," he said, "is that they are public and anyone can download them." He says that he stays away from other types of molecular markers, such AFLPs, because they are often patented. Yet, when I asked about certain patents on methods for using SSRs for particular purposes, such as SCN resistance, he admitted that he still had to look into it.

Part of the reason why SCN is such a serious disease problem in Canada is that the varieties in use here are generally susceptible. According to the plant breeder from the University of Guelph, there is not enough variation in the genetic resistance of current plant varieties.[39] "There are not enough resistance genes identified and old sources are not good anymore." He explained that diversity in genetic resistance to other diseases, such as Phytophthora, has been key in keeping these disease levels down over the years. One of the plant breeders with AAFC, who was involved in the development of the first early-maturing lines, credits this diversity to the extensive use of "Swedish germplasm," which gives Canadian short-season varieties "a wider base than the genetic base of soybean in the Midwest."[40]

Overall, however, the North American soybean crop is remarkably uniform. In a 1983 study of the pedigree of 158 commercial soybean cultivars from the U.S. and Canada, Delannay concluded that ten soybean lines provided 80 percent of the genetics within the North of American soybean crop.[41] According to the U.S. plant breeder, "Fewer than a hundred cultivars have contributed as much as a single gene to the U.S. soybean crops."[42] He and his team have been leading efforts to diversify the crop in the U.S., but with only limited success. In the 1980s, they began a co-operative breeding effort with twenty other programs, both public and private, to introduce "exotic germplasm" into the high-yielding U.S. soybean cultivars. But by the early 1990s all of the private breeding programs dropped out over concerns with patent protection and the time that it was taking to bring varieties to market. According to him, the private programs "felt that they couldn't justify the resources. Private breeding is incredibly competitive and it's difficult to justify spending resources when there is not a good chance of producing a variety in the first cycle of breeding."

He was concerned that private programs are limiting the diversity in the plant material they work with. He said that there is amazingly little overlap between the soybean cultivars of different companies, which shows that private breeding programs are almost exclusively looking within their own collections. And private breeding programs are not letting their material get into public programs either, leaving the public sector, with its much smaller R & D budgets, to work with lower-yielding varieties that do not have much of a chance on the seed market.

One of the AAFC plant breeders told me he is worried that a similar situation is unfolding in Canada.[43] He said that some private breeders do exchange with each other and with public programs, but large seed firms tend to work only within their own collections. As a result, "you get one pool of germplasm developing for the companies and another pool for the public breeders. . . . Companies generally think they can do it on their own." He said that this will inevitably push public programs out of soybean breeding. As he put it, the fundamental issue is "whether we think it is necessary to have public breeding programs around."

Another AAFC soybean breeder said that in Canada there is still a fair amount of exchange between the public and private soybean breeding programs.[44] Similarly, a breeder with the University of Guelph says that he has yet to come across a private breeder who is unwilling to share, and he says his program regularly exchanges varieties with private breeders.[45] Yet, the more experienced AAFC breeder, who has followed the development of the Canadian soybean seed market since its early days, pointed out that in the 1990s companies began insisting on material transfer agreements, which prohibited selection from within, or back-crossing with, their patented varieties.[46]

All three breeders also insisted that the open relationship with the private sector does not exist when it comes to GM crops. According to the more senior AAFC breeder, when the private sector became involved in short-season soybeans in the 1980s there was initially a very open environment of exchange among public and private programs. But "that has changed and is changing." Change, as he explained it, began in the late 1980s, "certainly with anything related to biotech," and he experienced the impacts directly.

Both AAFC plant breeders told me about how they had been trying to develop soybeans genetically modified for resistance to white mould, a disease against which breeders have had very little success in developing resistance. Using a gene that was developed by a scientist at the University of Toronto, who had published, not patented, his results, they were able to develop GM soybean plants with a high level of resistance to the disease. They then began discussions over the development of commercial varieties with Monsanto, which possessed certain patents involved in the genetic modification process, and the financial means to carry it through the regulatory process. It was at this point in their research that they learned that a French company had a patent on the gene they were using. In the end, Monsanto and the French company could not come to terms, leaving the AAFC breeders with no alternative but to abandon their research. One of the AAFC breeders explained to me that public programs have no other choice but to negotiate with large private firms when it comes to the development of GM crops. As he put it,

> Even if we had "freedom to operate," we wouldn't have the where-withal to do it. To take a GM crop from the initial stages of research to commercialization is a huge undertaking and the government is not going to give the millions it would cost to take the GM, white mould-resistant soybean technology through the regulatory process. We knew we didn't have complete "freedom to operate," but we thought that if we took it far enough then a company would pick it up."[47]

As in the U.S., the trend towards more and tighter intellectual property rights has spilled over into public institutions. One of the plant breeders with AAFC said that Canadian universities are adopting some of the same "concepts" on IPRs as their counterparts in the U.S., "trying to capture value for their research," and he admitted that "there's some pressure to do this within Agriculture Canada as well."[48] A plant breeder with the University of Guelph told me that he is "not a big fan of patents in breeding," but with patents on varieties granted in the U.S., "it has become a little naïve to rely on research exemptions" available in Canada.[49] While he would prefer not to patent genes, he acknowledged that he could be asked to do so in a "mild or less mild form" by university administrators, and he expected that he will likely be forced to patent any resistance genes he identifies. The long-term scenario, however, is that with the public sector acting like the private sector, there will be little justification for maintaining public funding. One of the AAFC plant breeders described how AAFC is already de-emphasizing conventional, oilseed soybean breeding and shifting its focus towards small, value-added markets, such as the Natto soybeans sold for export to Japan.[50] He said the federal government is pulling back from conventional breeding and other areas where private companies can do the work. For him, "There is always a pressure to get the public sector out of variety development because the private sector is there."

All three breeders resent these changes. According to one from AAFC:

> I really dislike this ability of doing one cross, making some selection, and then totally withdrawing the results from the rest of the community. Plant breeding is all incrementalism. We all stand on the shoulders of everyone who has gone before us and add our little bit. But our little bit, when you look at the whole contribution, is just a little bit, no grounds for claiming that as the final contribution that makes it more valuable than anybody else's contribution and allows you to get IPRs that preclude the continued use.[51]

When it comes to seed-saving, he dismissed the notion that it harms the seed supply. "Soybeans are pure-breeding and as long as you're not mixing you can easily save a variety from year to year, for as many generations as you want. It's really a matter of what exemptions do you allow. It's a society decision. Do you allow farmers to save their own seeds or not?"

The other plant breeder from AAFC said he believes that these changes will have a major impact on the direction of soybean breeding in Canada. He said, "You need a certain lack of pressure to do what we did with short-season soybeans, where there was no established market. No company would have invested the years and resources it took to put all the pieces together." According to him:

> With new seed varieties, most of the money is made during the first two to three years. Super varieties such as Maple and OAC Bayfield are rare. If you average it out over a long period of time, the new varieties don't do much better than the old. If a private company developed a super variety it would try to replace it earlier or make it Roundup Ready. An important advantage of making

varieties Roundup Ready is that it prevents brown-bagging[52] and gives breeders a way to replace varieties more quickly. You just take an existing variety and reintroduce it as Roundup. Public breeding programs, therefore, have an important role to play in making good, non-GM varieties available. . . . Initially we had a lot of hope for genetic engineering. We forgot about globalization and transnational companies. The fun has gone out of that research. There is a general feeling that companies can do everything. But the trouble is that the private sector can't and does not want to do everything. They are very focused on areas where everything clicks, where there is an established market.[53]

A germplasm developer with AAFC expresses a similar view.[54] He says that the private sector has been very good at riding upon the back of the long-term research efforts of the public sector, turning public parental lines into commercial varieties, but they are rarely interested in carrying out this long-term research themselves. They do not have the capacity to deal with more complex breeding problems and they have a natural tendency to ignore areas of research that might be good for farmers and the public, but could jeopardize sales of their other products, such as fertilizers and pesticides.

He worries about the increasing focus on high-tech genetic breeding work being driven by the private sector. "One tends to get blurred vision by looking only inside modern laboratories without spending enough time with field research and with older style science," he said. "High tech methods do have their potential applications. But goals like energy efficiency, mineral use efficiency, competitiveness against weeds are genetically complex, which makes them in good part (but not totally) unfit for current marker and DNA style approaches."

Breeding, as all of these public sector breeders agreed, is always about priorities. When I asked about reports suggesting that Roundup Ready varieties may have lower disease resistance, one AAFC breeder suggested that it probably "has to do with the number of breeding objectives that you can handle at one time. If you want to add in more breeding objectives it's going to be more difficult to develop disease resistance."[55]

For the plant breeder at the University of Guelph, patents make his work cumbersome and create more paperwork. "It is a lot more complicated than it was a few years ago when everyone was benefiting and everyone producing good lines," he says. The private sector programs are also suffering from patents, especially small seed companies. "If a small seed company wants access to a gene patented by a large seed company, it has nothing to offer the large seed company to gain access. The large company will not share unless there is an incentive. Eventually, one company will have a monopoly over all the good varieties." Backing this point up, the germplasm developer with AAFC said that in 1998 he put in a request to Monsanto for access to a Brazilian soybean cultivar for which Monsanto had purchased the rights from Brazil's national public research system, and he has yet to receive an answer. He has had similar frustrations trying to access a wheat variety from the private sector.[56]

For this reason, one of the AAFC breeders said that farmers should be concerned about the withdrawal of the public sector from conventional breeding. "The public sector provides a range of choice that won't be available if it's strictly private sector, especially as Roundup Ready and similar technologies predominate. If you want the conventional choice, it may be that private companies develop them and maybe they won't."[57] But it's not just about keeping the non-GM option open. According to the breeder with the University of Guelph: "It is important to keep public breeding programs in place because public programs pursue research that the private sector would not pursue. For example, crosses with wild soybean are extremely difficult to work with, but justified by the final aim. Companies focused on short-term results would not be able to carry this out." Of course, it is not only Canadian farmers who should be concerned about the disappearance of the conventional seed option and the focus on short-term results. These are matters with deep social and environmental implications for the entire society. The challenge now is to find ways to broaden and deepen participation in the practice and politics of Canada's seed system. Unfortunately, though, Canada's public breeding programs, with their top-down structures and closed cultures, are not well equipped for the task. Change is more likely to emerge from outside through the organized demands of Canadians wanting to salvage what is left of public plant breeding from the grip of privatization, and, allying with those breeders who still believe in civil service, to build a new seed system to serve the needs of all Canadians.

Chapter 8

Harvest

In February 2005 a team of Canadian nego-
tiators was in Bangkok for a meeting of the Convention on Biological Diversity. Their orders from Ottawa were to take the lead
in sabotaging a *de facto* moratorium on Terminator seeds. This
technology, patented and developed by the U.S. government and
the world's largest seed corporations, makes seeds sterile. It does
not increase yields, protect against pests, or offer any other benefit to farmers, or consumers for that matter. It simply prevents
people from saving seeds from their crops and replanting them.
The ETC Group estimates that if Canadian farmers were forced
to buy Terminator wheat seeds, it would cost them an additional
U.S. $85 million per year. Terminator is the ultimate weapon in
the commodification of the seed, the most direct means of forcing farmers to go back to the corporations every year for their
seed, and the Canadian government is on centre stage promoting it.

As I tried explaining the Terminator technology to my grandfa-
ther, a retired Saskatchewan farmer who sold the family farm in
the early 1980s, his face turned from a blank look of incompre-
hension to one of anger. He couldn't understand why anyone
would do such a thing to seeds and how such a basic part of farming that every farmer took for granted could be undermined in
such a short time. Why, he asked, would the Canadian government ever want to promote a technology that is so obviously detri-
mental to its farmers?

I suspect that the federal government sees no such contradiction. The bureaucrats would have various justifications: Terminator technology can prevent GM contamination, or it can encourage more investment in plant breeding by protecting a company's research. The truth is that there are many aspects of Canadian agricultural policy that are equally distasteful, yet easily justifiable within the logic of industrial agriculture. Take the case of migrant labourers on Canadian farms.

In April 2005, shortly after the Terminator scandal, migrant farm workers were in the headlines of Canada's major dailies. Quebec's Human Rights Tribunal had just issued a report condemning one of Canada's largest commercial vegetable farms, located about forty minutes southwest of Montreal, for segregating black workers, hired to pick and process vegetables, to a "blacks only" cafeteria on the farm – and as if that weren't enough, the room lacked heat, running water, proper toilets, refrigeration, and many other necessities. The workers also faced repeated verbal and physical abuse. For enduring the hard labour, racism, and daily bus ride to and from Montreal, the workers earned a measly $350 a week.[1] The media's coverage tended to portray the story as an anomaly, a throwback to a bygone age. Segregation, however, is becoming the norm in rural communities where there is migrant labour. Local people do not mix with the farm workers, who are often housed outside of town away from community activity, and who typically do not speak much French or English. The workers can not lay down any roots in Canada because their families have to stay in their home countries and they have no possibilities of acquiring citizenship. The fact that segregation takes place should not surprise anyone – it is the very essence of these "human resources" programs, which now supply tens of thousands of migrant workers to Canadian farms every year.[2]

Genetically engineered sterile seeds, racial segregation on farms: These are not aberrations in our food system; they have become necessary to its functioning. Industrial agriculture logically leads towards greater and greater exploitation of people and the environment. As we follow this path, the indefensible becomes defensible and, inevitably, the food system makes the whole of society sick.

Transformations at the level of the seed can lock us into an ever-greater industrialization of agriculture. What is happening at the level of the seed not only carries industrialization forward, it blocks out alternatives, and forces the food system towards greater corporate control. If we are to find a way out of this destructive quagmire we will have to address what is happening to the seed system and start to build anew.

From One Set of Elites to Another

Many people, unfortunately, still expect the government to abide by the framework of the bygone public seed system. For decades, policy was guided by the values of a productionist coalition that, for better or for worse, had the support of various sectors of the Canadian agricultural system. In actual practice, the government has moved on to another system in which, as seed industry spokesman Bill Leask has openly admitted, improvement is defined by a seed's "utility" to the corporate actors that control the global agri-food industry. These may be the seed/biotechnology/pesticide firms, the food processors and suppliers, the retailers or, as is increasingly the case, an alliance of all three.

The depth of the transformation in the Canadian seed system may not be readily apparent because the institutions of the productionist period, such as the Registration Review Committees and the public breeding programs, are still in place. However, they are or are becoming empty shells of what they once were. Public plant breeding programs are being reduced to a support role for the private seed sector. The institutions surrounding the variety registration system that once offered at least a small window for farmers and the general public to have input into the decision-making process, are gradually losing their influence and, in some cases, their legitimacy as spaces for public engagement. Decision-making used to be dominated by "experts" from the public sector, typically scientists, who assumed that they knew what was in the best interests of farmers and the general public, and that their opinions mattered. These days are drawing to a close. Decision-making over seeds is fast becoming the exclusive domain of another elite – the transnational corporations that dominate the global seed industry.

Government and industry maintain that the changes to the seed system will make it more "market-driven." This is a poor expression for the actual situation. The new global agri-food order is controlled by a small number of increasingly integrated firms that seek to enhance their position in relation to other corporate players through ownership and control of technology, from seed to supermarket. In the world that these corporations are building, the "market" does not exist, except in a very limited manner on the supermarket shelves, where few people possess the means to enjoy the luxury of exercising real choice. The main concern for today's transnational seed company is not selling seeds so much as it is selling proprietary "value-added" traits through grower contracts or identity preservation systems. The added "value" reflects the values of the corporations, which are generally concerned with encouraging the use of industrial inputs (pesticides and fertilizers) and providing efficiencies to downstream industrial processes. In this system there is no room for social, environmental, or even agronomic considerations that impinge on corporate profits.

Towards a New Vision for Seed Systems

The seed system will continue to move towards a consolidation of the third seed regime unless a popular movement emerges capable of challenging it. It is important to keep in mind that the agricultural policies that came together in the post-World War II period were public policies, certainly influenced by the international context of the time, but primarily defined by the domestic political context. The productionist policies of the era, including those of the second seed regime, emerged through grassroots struggle and a somewhat unfortunate widespread faith in the ability of industrialization to achieve the greater good. Grassroots community organizing and political mobilization were also critical to the breakdown of the postwar agri-food order.

The context at the beginning of the twentieth-first century does provide some hope that a new broad-based coalition can emerge to redefine food systems in Canada and articulate an alternative to the seed system promoted by the seed industry and the federal government. There is, indeed, a growing backlash against industrial agriculture, taking different forms around the world. In the South, this backlash is led by peasants and fisherfolk who increasingly coalesce around the concept of food sovereignty. Their struggles could reshape the global agri-food order, with important consequences for the export agriculture policies of countries like Canada.

In Canada, as in other industrialized countries, resistance to industrial agriculture occurs primarily through individual actions. These actions, like buying from independent family farmers, are important because they sustain and encourage the growth of the networks, infrastructure, and practices of alternative food and agriculture systems. But, in the words of food and farming activist Cathy Holtslander, the movement needs to fly with both wings. Collective, political action is also required. "We need to understand where the political and economic forces of agri-business are bearing down and why they have targeted certain parts of the food system and what is at stake," she says. Holtslander urges Canadians to rally around the holistic concept of food sovereignty and its support of family farmers. "As environmentalists, as farmers and as people who eat, we need to recognize our common interests – and we need to intervene effectively," she says.[3]

The picture in Canada is mixed when it comes to recent, collective, political actions over seeds. There have been some significant victories for grassroots campaigns: the CFIA's withdrawal of amendments to the *PBR Act*, Monsanto backing down from introducing GM wheat; and Canada's high-profile embarrassment over the Terminator ban in Bangkok. I also think that some of the farmer initiatives to challenge seed corporations are creating important spaces for political action. The group of Manitoba farmers testing farm-saved hybrid canola seed, and the various local networks of gardeners and farmers rekindling interest in heritage varieties, while less overtly political, are nevertheless effective. But there have been some stunning defeats too, such as the Percy Schmeiser case, and the general saturation of Canadian food and agriculture with GM canola and soybeans, as well as a number of less visible manoeuvres by the government and the seed industry that could turn out to be more problematic, like the CFIA's proposal to "modernize" Canada's seed regulations.

Looking at the overall picture in Canada, the core problem is only getting worse: farmers are losing more and more control over seeds and corporate control is expanding. Even though seed-saving continues with certain crops, there are hardly any farmers left who engage in any form of selection or plant breeding. The brutal truth is that seed-saving practices, and the skills and knowledge that go with them, are rapidly declining at the very moment when they are most needed. Alternatives to industrial agriculture will not work, and certainly not over the long term, unless seeds are under the control of those who plant them.

There is an urgent need for action on two fronts. On the one hand, direct work with seeds has to be strengthened. More farmer- and gardener-led seed networks need to be built, and the existing ones need to be strengthened. These efforts often only survive because of the selfless dedication of a few committed people; they need more support and participation, particularly from youth, and greater engagement with Canada's recent immigrants – what food theorist Harriet Friedmann refers to as the "biocultural diaspora" because of the wealth of culture, knowledge, and seeds that newcomers bring, and the potential they have to revitalize and enrich food production, provision, preparation, and consumption in Canada. It is sadly ironic that some of the most gifted members of this biocultural diaspora, the migrant farm workers who leave their diverse farms every year to cultivate Canada's industrial monocultures, are so excluded from such collective activity.

There is also a need for more research into developing varieties suited to the demands of non-industrial farms and local food systems. Here, Canadians can insist that their public breeding programs make more of a contribution. But, if public and even private breeding programs are to play a role, then care must be taken to avoid the top-down model of plant breeding that characterized the breeding programs of earlier years. Care must also be taken to ensure that the conditions for the commodification of the seed are not recreated within the alternative seed networks. The major seed companies are eyeing the organic seed market, and there is no reason why the same problems of corporate control could not be easily transposed to organic seeds. More fundamentally, if these seed networks are to become more than a marginal activity, the root causes that draw farmers into industrial agriculture and the technology treadmill need to be addressed. Seeds, food, and farming must not be viewed in isolation, but as inseparable parts of a new vision for an ecologically sound and socially just food system.

The second area in need of immediate attention concerns the subject matter of this book: the commodification of the seed. Corporations secure control over seeds by destroying the possibility for any alternatives. Through various regulatory and technological mechanisms they pre-empt the possible resurgence of farmer and gardener seed systems or, at least, force them underground into a precarious, illegal existence. Such laws, regulations, and technologies must be exposed and eliminated. Grassroots political action is necessary, not only around GM crops, where there is a fair amount of public awareness, but also around such obscure matters as variety registration, the Kernel Visual Distinguishability system, and plant breeders' rights. And we must approach these issues within a broader, collective vision for food and farming.

There are also opportunities for alliances with people trying to resist the expansion of corporate control into other sectors. The push for "intellectual property rights," for instance, extends well beyond agriculture. Just as patents, plant breeders' rights, and GM crops are being used by corporations to monopolize plant breeding and prevent farmers from saving seed, laws and technologies are being used to similar effect with pharmaceuticals, books, software, and music. In response, people are developing creative means to organize and resist. The free software and open source movements are directly challenging what they consider to be Microsoft's monopoly practices through their own approaches to program development and distribution. Music enthusiasts have set up peer-to-peer networks on the Internet, like the BitTorrent file distribution protocol. The creative commons community is promoting alternative forms of copyright to let authors put works in the public domain or minimize restrictions on what readers can do. Librarians are campaigning to save "fair use" principles, while AIDS activists throughout the world are demanding that medicine serve the health of people ahead of the bottom lines of pharmaceutical companies. And indigenous peoples continue to struggle against the intensifying theft and destruction of their knowledge and cultures.

Most of these efforts are not simply about resistance but about developing new, often community-focused means to produce and share books, music, software, or agricultural innovations. The idea of "hacking," which is at the core of the free software movement, can easily be applied to seed systems.[4] When farmers or gardeners select certain seeds or cross one plant with another, they are essentially hacking a new plant from the old, just as a software developer hacks a new application or modification into an existing computer program. The more people there are hacking, the more people there are developing and sharing the requisite skills and knowledge, and the more innovation there is taking place. In the end what you get is something much more suited to people's needs, reflecting the collective and open processes at work. The free software movement is a direct example of how effective hacking can be, as are the local seed systems that groups such as the Nayakrishi farmers in Bangladesh or the Réseau Semences Paysannes in France are revitalizing.[5]

There have been few attempts in Canada to try to bring these different movements together, despite the obvious commonalities. If these movements do converge, if linkages are created for solidarity and joint actions, a strong mass movement could emerge to articulate and advance an alternative perspective on innovation. We need to challenge the mindset that dominates Canadian seed policy, and policies in many other sectors.

Convergence, in the case of both of these potential coalitions, should not be confused with centralization. Centralization and monopoly control, embedded in both the public and corporate seed systems, are the bases of industrial agriculture. We must seek the opposite: food systems that thrive on diversity, on many people innovating collectively in many locations, supported by seed systems that are diverse, open, and free. Seeds must not be seen as mere industrial commodities.

Building this diversity presents a major practical and political challenge, especially in light of the seed industry's organizational unity and strength. But as the Canadian seed system comes further under corporate control, the space for alternatives will diminish and it will become even more difficult to contest the advance of industrial agriculture. There is an immediate danger that these transformations to the Canadian seed system will lock the country into a single food and agriculture system designed to meet the needs of a few transnational corporations. For those ill-served by this model, there really is no alternative but to rise to the challenge.

Notes

Chapter 1: Transformation

1 R.S. Anderson, E. Levy, and B.M. Morrison, *Rice Science and Development Politics: Research Strategies and IRRI's Technologies Confront Asian Diversity (1950–1980)* (Oxford: Clarendon Press, 1991); Bernhard Glaeser, ed., *The Green Revolution Revisited: Critiques and Alternatives* (London: Allen & Unwin, 1987).

2 GRAIN, "Editorial on food sovereignty," *Seedling*, April 2005 ⟨www.grain.org⟩.

3 Hugh Grant, Chairman and President of Monsanto, "Presentation to the Lehman Brothers chemicals conference," 20 March 2006.

4 Devlin Kuyek, "Lords of poison: The pesticide cartel," *Seedling*, June 2001.

5 Ibid.; *Agrow Reports* (322), 12 February 1999; Richard Lewontin, "The maturing of capitalist agriculture: Farmer as proletarian," in Fred Magdoff et al., eds., *Hungry for Profit: The Agribusiness Threat to Farmers, Food, and the Environment* (New York: Monthly Review Press, 2000).

6 Clive James, "Preview: Global status of commercialized transgenic crops: 2003," *ISAAA Briefs* (30) (Ithaca, NY: ISAAA, 2003); Charles M. Benbrook, "An appraisal of EPA's assessment of the benefits of Bt crops," Union of Concerned Scientists, 17 October 2000.

7 Henk Hobbelink, *Biotechnology and the Future of World Agriculture: The Fourth Resource* (New Jersey: Zed Books, 1991).

8 *Farm Journal*, October 1996, p.19. Such consolidation is discussed in D. Harhoff, P. Régibeau, and K. Rockett, "Some simple economics of GM food," *Economic Policy*, October 2001; M. Boehlje, "Industrialization of agriculture: What are the implications?" *Choices* (First Quarter), 1996; and William Heffernan, "Concentration of ownership and control in agriculture" in Magdoff et al., *op. cit.*

9 ETC Group, "World's Top 10 Seed Companies – 2006," 30 April 2007 ⟨www.etcgroup.org⟩; ETC Group, "Oligopoly Inc.: Concentration in corporate power, 2003," *ETC Communiqué* 82, 2003.

10 GRAIN, "Seed laws: Imposing agricultural apartheid," *Seedling*, July 2005 ⟨www.grain.org⟩. The data for Australia and Canada come from the International Seed Federation survey, cited in Bill Leask, "TUA's and their impact on IPRs; Presentation to the public institutions and management of intellectual property rights," Toronto, Ontario, 14 December 2005 ⟨www.ag-innovation.usask.ca⟩.

Chapter 2: Industrialization

1 Carlos Casares, "Entretien avec Gustavo Grobocopatel – La révolution agricole argentine peut nourrir le monde," *Agence France Presse*, 4 June 2007.

2 National Farmers' Union, "The farm crisis, bigger farms, and the myths of 'competition' and 'efficiency,'" Saskatoon, Saskatchewan, 20 November 2003.

3 Richard Levins and Richard Lewontin, *The Dialectical Biologist* (Cambridge, MA: Harvard UP, 1985).

4 D. Goodman, B. Sorj, and J. Wilkinson, *From Farming to Biotechnology: A Theory of Agro-Industrial Development* (Oxford: Basil Blackwell, 1987); Jack Ralph Kloppenburg Jr., *First the Seed: The Political Economy of Plant Biotechnology, 1492–2000* (New York: Cambridge UP, 1988).

5 Willard W. Cochrane, *The Development of American Agriculture: A Historic Analysis* (Minneapolis, MN: U. of Minnesota Press, 1979).

6 Harriet Friedmann, "The political economy of food: A global crisis," *New Left Review* (197), 1993; Heffernan, "Concentration of ownership"; Bruno Jean, *Agriculture et développement dans l'Est du Québec* (Sillery, QC: Presse de l'Université du Québec, 1985); Robin Pistorius and Jeroen van Wijk, *The Exploitation of Plant Genetic Information: Political Strategies in Crop Development* (New York: CABI Publishing, 1999); Anthony Winson, *The Intimate Commodity: Food and the Development of the Agro-Industrial Complex in Canada* (Toronto: Garamond Press, 1992).

7 P.H. Goering, Helena Norberg-Hodge, and J. Page, *From the Ground Up: Rethinking Industrial Agriculture* (London: Zed Books, 1993).

8 F.H. Buttel, "Ever since hightower: The new politics of agricultural research activism in the molecular age," paper prepared for presentation at the annual meeting of the American Sociological Association, Atlanta, 16 August 2003 〈www.agribusinessaccountability.org/page/256/1〉.

Chapter 3: Germination

1 Today, in the face of enormous obstacles and with little or no outside support, some First Nations people in Canada continue to care for traditional varieties passed down through generations, such as corn in Kahnawake, Quebec, and Saskatoon berries in the Locatee Lands of British Columbia. See Sean Robertson, "A place to re/member: Place, culture and resistance in Penticton, British Columbia," published by the Forum on Privatization and the Public Domain, 2005 〈forumonpublicdomain.ca/?q=node/47〉. Some early history of integration with settler agriculture can be found in Claude Aubé, *Chronologie du développement alimentaire au Québec* (Saint-Jean-sur-Richelieu, QC: Les Éditions du Monde Alimentaire, 1996); A.W. Crosby, *Ecological Imperialism: The Biological Expansion of Europe, 900–1900* (Cambridge: Cambridge UP, 1986); and Bruce Trigger, *The Huron: Farmers of the North* (Montreal: Harcourt Brace Jovanovich College Publishers, 1990.

2 Catharine Parr Traill, *The Canadian Settler's Guide* (Toronto: Times, 1857).
3 Agriculture and Agri-food Canada, *Canada Agriculture – The First Hundred Years* (Ottawa: Government of Canada, 1967) ⟨collections.ic.gc.ca/agrican/pubweb/hs1_toc.asp⟩.
4 T.H. Anstey, *Cent moissons: Direction générale de la recherche Agriculture Canada, 1886–1986* (Ottawa: Centre d'édition du gouvernement du Canada, 1986).
5 Stephan Symko, "From a single seed: Tracing the Marquis wheat success story in Canada to its roots in the Ukraine," a web publication of Research Branch, Agriculture and Agri-Food Canada, Ottawa, 1999.

Chapter 4: Nation

1 Symko, "From a single seed."
2 Jean-Pierre Berlan, "La génétique agricole : 150 ans de mystifications. Des origines aux chimères génétiques," in Jean-Pierre Berlan, *La guerre au vivant : OGM et mystifications scientifiques* (Montréal: Comeau & Nadeau, 2001); Raoul A Robinson, *Return to Resistance: Breeding Crops to Reduce Pesticide Dependence* (Ottawa: IDRC, 1996).
3 Kloppenburg, *First the Seed*; Pistorius and Wijk, *The Exploitation of Plant Genetic Information*.
4 Cited in Kloppenburg, *op. cit.*, p.69.
5 Under the Mendelian approach, scientists collected plant varieties that possessed genes for traits of interest, from farmers' fields. They then transferred the desired genes to their established or "elite" cultivars, and finally sent these "finished" varieties out to farmers. Every once in a while a new variety would be released to replace the previous one. Farmers' fields thus became a contradictory site for both sourcing genetic resources and adopting elite cultivars, with the success of the latter assuring the erosion of the former. As the new Mendelian varieties were sent out far and wide, the agricultural diversity in farmers' fields, essential to the Mendelian breeding efforts, declined rapidly. This paradox did not inspire fundamental questions about the logic of such an approach; instead, it became a justification for further intervention. Gene banks, curated and controlled by scientists, were set up to preserve the diversity that was rapidly disappearing from farmer's fields. In this manner, public scientists brought a high degree of control over plant breeding decisions into their own hands or, more accurately, within the bureaucracy of the state.
6 Anstey, *Cent moissons : Direction générale de la recherché Agriculture Canada, 1886–1986;* Symko, "From a single seed."
7 J.B. Campbell, "The Swift Current research station 1920–70" (Ottawa: Government of Canada, 1971); V.C. Fowke, *The National Policy and the Wheat Economy* (Toronto: University of Toronto Press, 1957).
8 C.F. Wilson, *A Century of Canadian Grain: Government Policy to 1951* (Saskatoon: Western Producer Prairie Books, 1978).
9 Symko, *op. cit.*

10 Anstey, *Cent moissons.*

11 Western farmers not practising intensive wheat monoculture, such as the Mennonites of Manitoba, were comparatively much better off during these difficult years. See Campbell, *op. cit.*

12 Friedmann, "The political economy of food"; Éric Montpetit and William D. Coleman, "Policy communities and policy divergence: Agro-environmental policy development in Quebec and Ontario," *Canadian Journal of Political Science* 32(4), 1999; G. Marchildon, "Canadian-American agricultural trade relations; A brief history," in G. Marchildon, ed., *Agriculture at the Border: Canada-U.S. Trade Relations in the Global Food Regime* (Regina: Canadian Plains Research Center, 2000), p.35; Winson, *op. cit.*

13 Fowke, *op. cit.*; Winson, *op. cit.*

14 Statistics Canada, Agriculture Division, "M525. Expenses for fertilizer, Canada, 1926 to 1976 ⟨www.statcan.ca/english/freepub/11-516-XIE/sectionm/sectionm.htm#M525⟩.

15 Winson, *op. cit.*, p.65.

16 Jean, *Agriculture et développement.*

17 Buttel, "Ever since Hightower"; Friedmann, "The political economy of food"; P. Thompson, "The reshaping of conventional farming: A North American perspective," *Journal of Agricultural and Environmental Ethics* 14(2), 2001.

18 Robert L. Zimdahl, "Rethinking agricultural research roles," *Agriculture and Human values* 15(1), 1998.

19 "Interview with germplasm developer," Agriculture and Agri-food Canada, 20 March 2003.

20 A corn plant reproduces by way of cross-pollination with another corn plant, whereas wheat generally self-pollinates. In self-pollinating crops the frequency of cross-pollination is low.

21 Berlan, "La génétique agricole."

22 T.W. Bruulsema, M. Tollenaar, and J.R Heckman, "Boosting crop yields in the next century," *Better Crops* 84 (1), 2000, pp. 9–13.

23 R.M.A. Loyns and A.J. Begleiter, "An examination of the potential economic effects of plant breeders' rights on Canada," Working Paper for Consumer and Corporate Affairs Canada, 1984, p. 109.

24 R.K. Downey, "Presentation on plant breeders' rights," *Proceedings of the Conference on Plant Breeding and Breeders' Rights in Canada*, Crop Science Department, University of Guelph, 15–16 June 1971.

25 A.E. Hannah, Assistant Director General of the Research Branch of Agriculture Canada, cited in *Proceedings of the Conference on Plant Breeding.*

26 Brewster Kneen, *The Rape of Canola* (Toronto: NC Press, 1992).

27 Pamela Cooper, "Plant breeders' rights: Some economic considerations, a preliminary report," Economic Working Paper, Agriculture Canada, 1984, p.24.

28 Loyns and Begleiter, "An examination of the potential economic effects," p.21.

29 R. Gray, S. Malla, and S. Ferguson, "Agriculture research policy for crop improvement in Western Canada: Past experience and future directions,"

report prepared for Saskatchewan Agriculture and Food, February 2001, p.6.

30 *The Seeds Act* was amended in 1937 to exclude vegetables, with the exception of potatoes, from registration.

31 Anstey, *Cent moissons.*

32 W.T. Bradnock, "Plant breeders' rights and plant patents," *Proceedings of the Workshop on Plant Gene Patenting* (Ottawa: CARC, 1987).

33 Kneen, *The Rape of Canola.*

34 Ibid., p.37.

35 Ibid., p.32.

36 Ibid., p.31.

37 Ibid., p.60.

Chapter 5: Corporation

1 Montpetit and Coleman, "Policy Communities."

2 Graham Riches, "Advancing the human right to food in Canada: Social policy and the politics of hunger, welfare, and food security," *Agriculture and Human Values* 16(2), 1999.

3 Kenneth Dahlberg, "Democratizing society and food systems: Or how do we transform modern structures of power?" *Agriculture and Human Values* 18 (2), 2001; M. Koc and K.A. Dahlberg "The restructuring of food systems: Trends, research and policy issues," *Agriculture and Human Values* 16(2), 1999; Thompson, "The reshaping of conventional farming"; E. Wall and B. Beardwood, "Standardizing globally, responding locally: The new infrastructure, ISO 14000, and Canadian agriculture," *Studies in Political Economy* (64), 2001.

4 Friedmann, "The political economy of food."

5 G. Galizzi and L. Venturini, eds., *Vertical Relationships and Coordination in the Food System* (New York: Physica-Verlag, 1999)

6 Winson, *The Intimate Commmodity*, pp. 159–166.

7 Boehlje, "Industrialization of agriculture: What are the implications?"; Kuyek, "Lords of Poison"; Lewontin, "The maturing of capitalist agriculture."

8 R.E. Goodhue and G.C. Rausser, "Value differentiation in agriculture: Driving forces and complementaries," in Galizzi and Venturini, *op. cit.*

9 Caroline Daniel, "The Cargill approach: 'We don't lobby. We go and share information,'" *Financial Times*, 26 February 2003.

10 A.M. Azzam, "Vertical relationships: Economic theory and empirical evidence," in Galizzi and Venturini, *op. cit.*; M. Boehlje and S. Sonka, "Structural realignment in agriculture: How do we analyze and understand it?" University of Illinois, 1998 ⟨www.ag.uiuc.edu/famc/program98/sonka.htm⟩.

11 Bill Vorley, "Food Inc.: Corporate concentration from farm to consumer" (UK Food Group: London, 2003).

12 Wendy Larmer, "Neo-liberalism: Policy, ideology, governmentality," *Studies in Political Economy* (63), 2000.

13 Rodney Loeppky, "Gene Production: A Political economy of human genome research," *Studies in Political Economy* (60), 1999; John Portz and Peter Eisinger, "Biotechnology and economic development: the role of the states," *Politics and the Life Sciences* 9(2), 1991.

14 Martin Kenney, *Biotechnology: The University-Industrial Complex* (New Haven: Yale UP, 1986).

15 Murray Moo-Young and Lamptey, eds., *Proceedings of Biotechnology Day II*, University of Waterloo, 6 November 1984.

16 National Biotechnology Advisory Committee, "Report from the national biotechnology advisory committee" (Ottawa: Government of Canada, 1984), p. v.

17 Devlin Kuyek, *The Real Board of Directors: The Construction of Biotechnology Policy in Canada, 1980–2002* (Sorrento, BC: Ram's Horn, 2002); Greg Pichler, "The technological capability of Canada, Inc.," *Management Science* (232), 1989.

18 National Biotechnology Advisory Committee, "Fifth report of the National Biotechnology Advisory Committee," (Ottawa: Government of Canada, 1991).

19 James Heller, "Background economic study of the Canadian biotechnology industry," Paper Commissioned by Industry Canada and Environment Canada, 1992.

20 John Vose, former director of National Research Council's Industrial Research Assistance Program, cited in Moo-Young and Lamptey, *op. cit.*

21 W.T. Stanbury, "Reforming the federal regulatory process in Canada, 1971–1992," *Annex to the Report of the Sub-Committee on Regulations and Competitiveness of The Standing Committee on Finance*, Ottawa, 10 December 1992.

22 Ibid., p.69.

23 Cited in Kuyek, *The Real Board of Directors*, p.35.

Chapter 6: Commodification

1 ETC Group and Greenpeace, "Who owns terminator patents?" Ban Terminator website ⟨www.banterminator.org⟩; ETC Group, "Terminator technology: Five years later," *ETC Communiqué* 79, 2003.

2 Vic Duy, "A brief history of the Canadian patent system," prepared for the Canadian Biotechnology Advisory Committee, January 2001 ⟨cbac-cccb.ca/epic/internet/incbac-cccb.nsf/vwGeneratedInterE/h_ah00128e.html⟩; Michelle Swenarchuk, "The Harvard mouse and all that: Life patents in Canada," Canadian Environmental Law Association, October, 2003.

3 Judge J. MacKay, "Judgement in the case of Monsanto Canada Inc. and Monsanto Inc. versus Percy Schmeiser and Schmeiser Enterprises Ltd.," Federal Court of Canada, 29 March 2001.

4 Lyle Friesen et al., "Evidence of contamination of pedigreed canola (B. napus) seedlots in Western Canada with genetically engineered herbicide resistance traits," *Agronomy Journal* (95), 2003.

5 Kelly Hearn, "Don't cry to them, Argentina: Is Monsanto playing fast and loose with Roundup Ready soybeans in Argentina?" *GRIST Magazine*, 22 September 2006; Rachel Nellen-Stucky and François Meienberg, "Harvesting royalties for sowing dissent? Monsanto's campaign against Argentina's patent policy," trans. Maja Ruef, Berne Declaration, October 2006; and GRAIN, "Confronting contamination: 5 reasons to reject co-existence," *Against the Grain*, April 2004 ⟨www.grain.org⟩.

6 Peter Shinkle, "Monsanto reaps some anger with hard line on reusing seeds: Agriculture giant has won millions in suits against farmers," *St-Louis Post-Dispatch*, 12 May 2003.

7 Michael Mascarenhas and Lawrence Busch, "Seeds of change: Intellectual property rights, the technological treadmill, and the consequences for seed saving practices in the United States," presented at the Rural Sociological Society Annual General Meeting, Montreal, 27–9 July 2003.

8 BASF Canada, 2003 ⟨www.agsolutions.ca/pub/west/clearfield/commitment/gen.cgi/main⟩.

9 Personal communication, 23 February 2003.

10 C&M, 2003, ⟨www.redwheat.com/identity_preserved_program.htm⟩.

11 Laura Rance, "Canola heads for the big leagues" *Farmers' Independent Weekly*, 25 July 2002, p.14.

12 Pistorius and Wijk, *The Exploitation of Plant Genetic Information*.

13 Barry Wilson, "Industry forms alliance to help enforce seed rights," *Western Producer*, 4 December 1997.

14 Alberta Seed Industry, "PBR: They mean business," 1 September 2000 ⟨www.seed.ab.ca/pbr/sf20000901.shtml⟩; "40–50 people found in violation of *PBR Act* to date," *Germination*, September 2001.

15 Interview with Plant Breeder (6), Svalof Weibull, Saskatoon, Saskatchewan, 21 November 2002.

16 Robynne Anderson, "Mean what you say," *Germination*, January 2004.

17 Canadian Food Inspection Agency, *Ten-year Review of the Plant Breeders' Rights Act*, Government of Canada, 2003 ⟨www.inspection.gc.ca⟩.

18 Bill C-80: Part 10 – Plant Breeders Rights Act, 1998.

19 Canadian Seed Trade Association, "Canadian Seed Trade Association position paper on biotechnology," July 2001 ⟨www.cdnseed.org⟩.

20 Canadian Grain Commission, "Canadian Identity Preserved Recognition System," 2 June 2003 ⟨www.grainscanada.gc.ca/Prodser/ciprs/ciprs1-e.htm⟩.

21 Prairies Registration Recommending Committee for Grain, "PRRCG report: From the 2002 Prairies registration recommending committee for grain annual meeting," Meristem Land and Science, Spring 2002.

22 Meristem Land and Science, "The future of variety registration," 3 May 2002 ⟨www.meristem.com/prrcg/prrcg02.html⟩.

23 *Germination*, July 2002.

24 *SeCan News*, December 1996.

25 *SeCan News*, December 2000.

26 John Morriss, "Viewpoint," *Farmers Independent Weekly* 25 July 2002.

27 R.K. Downey, "Presentation on Plant Breeders' Rights."

28 Loyns and Begleiter, *op. cit.*

29 Berlan, "La génétique agricole."
30 Leask, "TUA's and their impact on IPRS."
31 Cooper, "Plant breeders' rights," p.24.
32 Danish Institute of Agricultural Sciences, "Proceedings of the 1st European conference on the co-existence of genetically modified crops with conventional and organic crops (GMCC-03): GM crops and co-existence," Danish Institute of Agricultural Sciences, Research Centre Flakkebjerg, 13–14 November 2003.
33 Reg Sherren, "The controversy over genetically modified canola." *CBC News and Current Affairs*, 21 March 2002.
34 Friesen et al., *op. cit.*
35 Laura Rance, "Farmers urged to ensure . . . ," *The Manitoba Co-operator*, 1 March 2001.
36 Organic Agriculture Protection Fund, "Organic farmers gain key piece of evidence in class action," Media Release, 26 June 2002.
37 Karen Charman, "Seeds of domination: Don't want GMOs in your food? It may already be too late," *In These Times*, 10 February 2003.
38 Margaret Mellon and Jane Rissler, "Gone to seed: Transgenic contaminants in the traditional seed supply," Union of Concerned Scientists, 2004.
39 Laura Rance, "Annual variety exams pose difficult questions," *The Manitoba Co-operator* 13 March 1997, p.16.
40 Canadian Grain Commission, "Identity preserved systems in the Canadian grain industry: A discussion paper," Government of Canada, December 1998.
41 Rance, "Farmers urged to ensure . . . "
42 Canadian Seed Trade Association, "Affidavit systems: A position paper of the Canadian Seed Trade Association," January 2003 〈cdnseed.org/press/Affidavitsystemsposition.pdf〉.
43 Mark Condon, "Seed genetic purity in the pre and post biotechnology eras," conference presentation, Pew Initiative and the Economic Research Service of the U.S. Department of Agriculture, Minneapolis, 11 September 2002.
44 Canadian Seed Institute 〈www.csi-ics.com/en〉.
45 Canadian Grain Commission, "Canadian identity preserved recognition system."
46 Terry Boehm, "Variety registration changes will have far-reaching consequences," op-ed released by the National Farmers Union, 8 November 2006.
47 Kneen, *The Rape of Canola*.
48 Jeff Guest, "Commentary," *Germination*, March 2003, p.36.
49 "Thinking outside the box," *Germination Magazine*, January 2007 〈www.germination.ca/pdfs/Germ_Jan07.pdf〉.
50 Ibid.
51 Personal communication with Bernard Estevez, Quebec agronomist, 26 June 2007.
52 Daniel Winters, "Seed agreement puts onus on grower," *Western Producer*, 25 January 2007.
53 Mary MacArthur, "Seed cleaners refuse to play police," *Western Producer*, 27 April 2006.

54 Interview with Terry Boehm, National Farmers Union, 25 June 2007.
55 Allan Dawson, "Big seed companies don't want to invest in competitors' research: Alberta's plan for farmer and publicly funded crop development," *Farmers' Independent Weekly*, December 2007.
56 Ian Bell, "Hybrid canola practices challenged," *Western Producer*, 31 May 2005.
57 Thiago Brandão and Raquel Mariano, "Embrapa ajudará na fiscalização de plantações de soja não certificada," *Agência Brasil*, 21 October 2006.
58 *Germination*, January 2004.

Chapter 7: Privatization

1 Moo-Young and Jonathan Lamptey, eds., *Proceedings of Biotechnology Day II*.
2 Cited in ibid.
3 Loyns and Begleiter, "An examination of the potential economic effects," pp. 41–2.
4 Downey, "Presentation on plant breeders' rights."
5 Cooper, "Plant breeders' rights."
6 Sean Pratt, "Agriculture loses grip on research money," *Western Producer*, 14 August 2007.
7 Gray et al., "Agriculture research policy," p.1.
8 Kneen, *The Rape of Canola*, pp. 37–8.
9 Interview with plant breeder (1), Agriculture and Agri-food Canada, 4 November, 2002.
10 Interview with plant breeder (5), University of Saskatchewan, 21 November 2002.
11 Anon., untitled, *Nature Biotechnology* 17 October 1999, p.936.
12 Interview with germplasm developer, Agriculture and Agri-food Canada, 20 March 2003.
13 Interview with plant breeder (5).
14 Personal communication with two plant breeders from the University of Guelph, February 2003.
15 Allan Dawson, "Concerns raised about royalties for public breeders," *Farmers' Independent Weekly*, 9 January 2003.
16 The patent number is U.S.6303849, for "Brassica juncea lines bearing endogenous edible oils."
17 Interview with plant breeder (7), Saskatchewan Wheat Pool, 21 November 2002.
18 Interview with plant breeder (5).
19 Gray et al., "Agriculture research policy."
20 Ibid., p.2.
21 Ibid., p.2.
22 Interview with plant breeder (8), University of Guelph, 13 February 2003.
23 Keith Whigham, "Soybean history," Dept of Agronomy, Iowa State University, 2003 ⟨www.agron.iastate.edu/soybean/history.html⟩.

24 Lester Brown, "The United States and China: The soybean connection," *Worldwatch Newsbriefs*, 9 November 1999 ⟨www.worldwatch.org/alerts/991109.html⟩.
25 James, "Preview: global status."
26 Lawrence Busch, "Lessons unlearned: How biotechnology is changing society," conference presentation, Seattle, WA, 1–3 June 2003; Mascarenhas and Busch, "Seeds of Change."
27 ETC Group, "Oligopoly Inc.: Concentration in corporate power, 2003"; Harhoff et al., "Some simple economics of GM food."
28 William Shurtleff and Akiko Aoyagi, *Soyabean in Canada: Bibiliography and Sourcebook, 1855–1993* (Lafayette, CA: Soyfoods Canada, 1993).
29 Anstey, *Cent moissons*; H.D. Voldeng, "Working with breeding short-season soybean in Canada (interview)," *SoyaScan Notes*, 2 March 1993.
30 Shurtleff and Aoyagi, *op. cit.*; J.A. Wrather, T.R. Anderson, D.M. Arsyad, Y. Tan, L.D. Ploper, A. Porta-Puglia, H.H. Ram, and J.T. Yorinori, "Soybean disease loss estimates for the top ten soybean-producing countries in 1998," *Canadian Journal of Plant Pathology* (23), 2001.
31 *Agrow Reports*, 19 January 2004.
32 Interview with plant breeder (2), Agriculture and Agri-food Canada, 4 November 2002.
33 Interview with plant breeder (1).
34 Interview with plant breeder (3), University of Guelph, 12 November 2002.
35 Interview with plant breeder (4) and germplasm curator, United States Department of Agriculture, 18 November 2002.
36 Ibid.
37 Interview with plant breeder (1).
38 Interview with plant breeder (3).
39 Ibid.
40 Interview with plant breeder (2).
41 Q. Yang and J. Wang, "Agronomic traits correlative analysis between interspecific and intraspecific soybean crosses," *Soybean Genetics Newsletter 27*, 10 April 2000 ⟨www.soygenetics.org/articles/sgn2000-003.html⟩.
42 Interview with plant breeder (4).
43 Interview with plant breeder (2).
44 Interview with plant breeder (1).
45 Interview with plant breeder (3).
46 Interview with plant breeder (2).
47 Ibid.
48 Interview with plant breeder (1).
49 Interview with plant breeder (3).
50 Interview with plant breeder (1).
51 Ibid.
52 Brown-bagging refers to the practice common to farmers of selling saved, non-certified seeds from their farms to other farmers.
53 Interview with plant breeder (2).
54 Interview with germplasm developer, Agriculture and Agri-food Canada, 26 June 2007.

55 Interview with plant breeder (1).
56 Interview with germplasm developer, Agriculture and Agri-food Canada, 26 June 2007.
57 Interview with plant breeder (1).

Chapter 8: Harvest

1 Martin Patriquin, "Quebec farm segregated black workers," *The Globe and Mail*, 30 April 2005, p. A2.
2 Sue Ferguson, "Hard time in Canadian fields: Conditions tough for Canada's migrant workers," *Maclean's*, 11 October 2004.
3 Cathy Holtslander, "Farmer-environmentalist solidarity: Resisting the agribusiness agenda," presentation to the Canadian Environment Network annual conference, Montreal, 20 October 2006.
4 Tim Jordan and Paul A. Taylor, *Hactivism and Cyberwars: Rebels with a Cause?* (New York: Routledge, 2004).
5 Réseau Semences Paysannes ⟨www.semencespaysannes.org⟩.

Interviews and Personal Communications

Interview with plant breeder (1), Agriculture and Agri-food Canada, Ottawa, 4 November 2002.

Interview with plant breeder (2), Agriculture and Agri-food Canada, Ottawa, 4 November 2002.

Telephone interview with plant breeder (3), University of Guelph, 12 November 2002.

Telephone interview plant breeder and germplasm curator (4), United States Department of Agriculture, 18 November 2002.

Interview with plant breeder (5), University of Saskatchewan, Saskatoon, 21 November 2002.

Interview with plant breeder (6), Svalof Weibull Company, Saskatoon, 21 November 2002.

Interview with plant breeder, Saskatchewan Wheat Pool, Saskatoon, 21 November 2002.

Interview with director, Plant Genetic Resources Centre, Agriculture and Agri-food Canada, Saskatoon, 22 November 2002.

Telephone interview with plant breeder, University of Guelph, 13 February 2003.

Email communication from Richard Gold, professor of law, McGill University, 23 February 2003.

Interview with germplasm developer, Agriculture and Agri-food Canada, Sainte-Foy, Quebec, March 20, 2003 and telephone interview, 26 June 2007.

Interview with Terry Boehm, National Farmers Union, 25 June 2007.

Interview with Bernard Estevez, agronomist, 26 June 2007.

Going Further

Books

Berlan, Jean-Pierre. *La guerre au vivant: OGM et mystifications scientifiques* (Montréal: Comeau & Nadeau, 2001).

Fowler, Carey and Pat Roy Mooney. *Shattering: Food Politics and the Loss of Genetic Diversity* (Tuscon, AZ: University of Arizona Press, 1990).

Hobbelink, Henk. *Biotechnology and the Future of World Agriculture: The Fourth Resource* (New Jersey: Zed Books, 1991).

Kenney, Martin. *Biotechnology: The University-Industrial Complex* (New Haven: Yale University Press, 1986).

Kloppenburg, Jack Ralph Jr. *First the Seed: The Political Economy of Plant Biotechnology, 1492–2000* (New York: Cambridge University Press, 1988).

Kneen, Brewster. *From Land to Mouth: Understanding the Food System* (Toronto: NC Press, 1989).

_____, *The Rape of Canola* (Toronto: NC Press, 1992).

_____, *Farmageddon: Food and the Culture of Biotechnology* (Gabriola Island, British Columbia: New Society Publishers, 1999).

Kuyek, Devlin. *The Real Board of Directors: The Construction of Biotechnology Policy in Canada, 1980–2002* (Sorrento, BC: Ram's Horn, 2002).

Patel, Raj. *Stuffed and Starved: Markets, Power and the Hidden Battle for the World Food System* (London: Portobello Books, 2007) ⟨www.stuffedandstarved.org⟩.

Pistorius, Robin and Jeroen van Wijk, *The Exploitation of Plant Genetic Information: Political Strategies in Crop Development* (New York: CABI Publishing, 1999).

Winson, Anthony. *The Intimate Commodity: Food and the Development of the Agro-Industrial Complex in Canada* (Toronto: Garamond Press, 1992).

Websites

Beyond Factory Farming ⟨www.beyond factoryfarming.org⟩.
Canadian Biotechnology Action Network ⟨www.cban.ca⟩.
ETC Group ⟨www.etcgroup.org⟩.
Forum on the Patenting of Life ⟨lists.riseup.net/www/info/fpl-fbv⟩.

Forum on Privatization and the Public Domain ⟨forumonpublicdomain.ca⟩.
GRAIN ⟨www.grain.org⟩.
National Farmers Union ⟨www.nfu.ca⟩.
Ram's Horn ⟨www.ramshorn.ca⟩.
Saskatchewan Organic Directorate ⟨www.saskorganic.com⟩.
Seeds of Diversity ⟨www.seeds.ca⟩.

Index